Liaison Functions

Core Programming Topic at the 2011 AIChE Annual Meeting

Minneapolis, Minnesota, USA
16-21 October 2011

ISBN: 978-1-61839-738-6

Printed from e-media with permission by:

Curran Associates, Inc.
57 Morehouse Lane
Red Hook, NY 12571

CURRAN ASSOCIATES INC.
proceedings
.com

Some format issues inherent in the e-media version may also appear in this print version.

Printed by Curran Associates, Inc. (2012)

AIChE
3 Park Avenue
New York, NY 10016-5991

Phone: (203) 702-7660
Fax: (203) 775-5177

www.aiche.org

Additional copies of this publication are available from:

Curran Associates, Inc.
57 Morehouse Lane
Red Hook, NY 12571 USA
Phone: 845-758-0400
Fax: 845-758-2634
Email: curran@proceedings.com
Web: www.proceedings.com

TABLE OF CONTENTS

EFFECTIVE APPLICATION OF PROJECT RISK MANAGEMENT

Stephen L. Cabano

Pathfinder, LLC
Cherry Hill, NJ

Biography

Steve Cabano is President of Pathfinder, LLC, a project management consulting firm specializing in the planning, development and execution of domestic and international capital plant construction projects. With more than 25 years of project engineering and project management experience in the industrial and commercial facilities industry, he has provided cost, scheduling, procurement, and similar project related services for owner and government clients in the Petroleum, Petrochemical, Chemical, Environmental, Pharmaceutical, Industrial and Commercial industries in the United States and throughout the world.

Introduction

In today's complex capital project environment, a project manager has a myriad of risks to consider and address. The traditional risks include: managing to the project cost and schedule commitments, addressing regulatory and environmental concerns, availability of internal resources vs. contracting out for services, among others. In today's increasingly active capital project execution environment, a whole new set of potential risks have arisen. Several of these include availability of competent engineering and construction contractors, experience of engineering and construction supervisors, material and equipment availability, the multi cultural nature of today workforce, etc.

Basis of Observations

This paper will address capital project risk management issues from an owner's perspective concerning projects in the process industry. For the purposes of this paper, we define the process industry as the oil, gas, chemical, petrochemical and related industry sectors. We will also discuss the application of risk management tools and techniques across an industry accepted, 5 Phase "Best Practice" Capital Project Delivery Process as outlined below in Figure 1:

EFFECTIVE APPLICATION OF PROJECT RISK MANAGEMENT

Figure 1

The Project Management Process

1 BUSINESS PLANNING	2 ALTERATIVES ANALYSIS	3 BASIC ENGINEERING	4 EXECUTION -EPC-	5 OPERATE & EVALUATE
OBJECTIVES: • Clearly frame goal. • Identify opportunities. • Test for Strategic fit with business objectives. • Preliminary assessment of uncertainties, potential return and associated risks. • Plan for next phase.	OBJECTIVES: • Generate alternatives. • Reduce uncertainty and quantify associated risks. • Develop expected value for selected alternatives. • Identify preferred alternative(s). • Plan for next phase.	OBJECTIVES: • Fully define scope. • Develop detailed execution plans. • Refine estimates & economic analysis to A/R level. • Confirm if expected value meets business objectives.	OBJECTIVES: • Implement Execution Plan. • Finalize Operating Plan • Collect, analyze, and share metrics & lessons learned.	OBJECTIVES: • Monitor performance. • Benchmark performance against objectives and competitors. • Share results and lessons learned. • Continue performance assessment and identify opportunities.

The Risk Management Process

The Project Management Institute (PMI) identifies Risk Management as one of the basic Project Management Bodies of Knowledge. (See Figure 2)

Figure 2

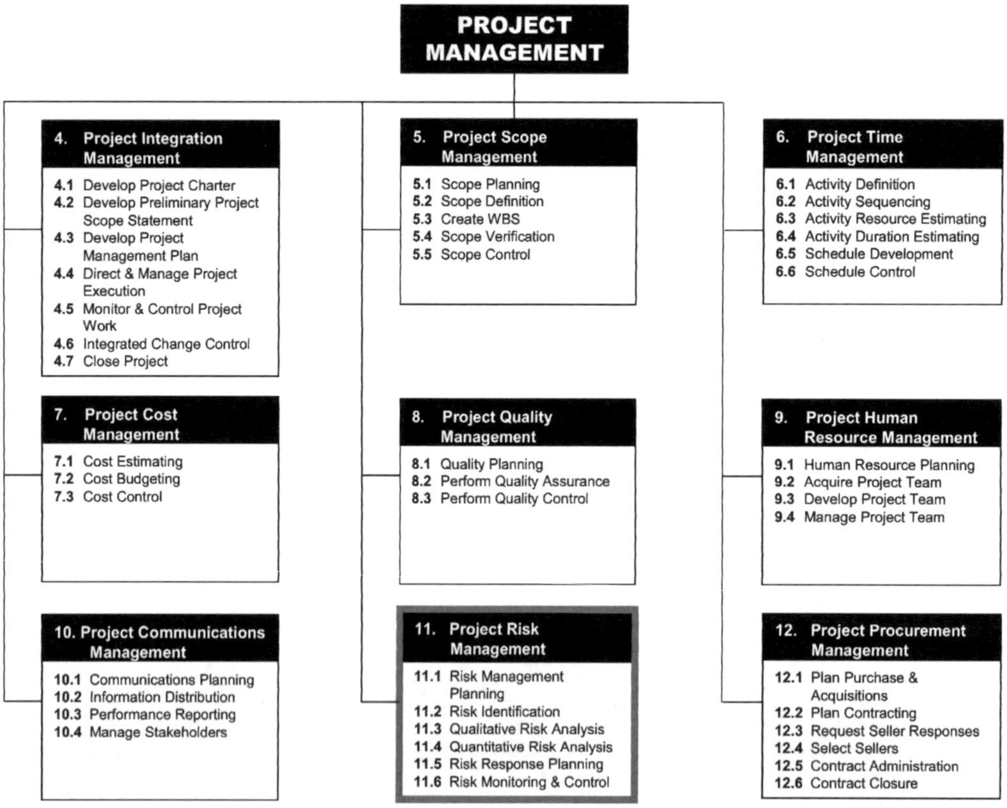

2008 PMBOK Guide

EFFECTIVE APPLICATION OF PROJECT RISK MANAGEMENT

PMI defines Risk Management in the following 6 categories:

- Risk Management Planning
- Risk Identification
- Qualitative Risk Analysis
- Quantitative Risk Analysis
- Risk Response Planning
- Risk Monitoring and Control

Let's take each one of these categories and identify the issues related to each in today's project execution environment.

Risk Management Planning

For a Risk Management approach to be effective there needs to be a company-wide established process which is applied, utilized and adhered to by all project representatives, as well as a project specific risk management plan integrated into the project's execution plan.

Company-wide processes should outline the corporate tools, resources, and deliverables by phase. A series of support "think lists"/risk considerations should be established for each project phase to assist the project team in identifying project risks. An example of this is outlined below in Figure 3:

Figure 3

Risk Management Think List by Project Phase

OPPORTUNITY (1)	ANALYSIS (2)	DEFINITION (3)	EXECUTION (4)	OPERATE (5)
• What are the business risks? • What market influences will impact this opportunity? • How will these risks impact the company?	• What are the opportunities and threats for each alternative evaluated? • How do these risks impact cost, schedule, operations, reliability, etc.? • What is the benefit of implementing a new technology? What are the risks?	• What are the project risks associated with the selected alternative? • What are the business risks associated with the selected opportunity? • How will the identified risks impact the overall project? • Are there mitigation options that should be incorporated into the base scope/execution plan?	• What new risks have been uncovered during detailed engineering, procurement, construction and Start-up? • How will these new risks impact the commitment made to management? • Are these late risks manageable and will they impact operations?	• What will impact the effective handover to operations? • What risks can be mitigated through application of effective Lessons Learned? • Are there risks that will affect operability/ maintainability and therefore operating costs and/or reliability/ utilization factors?

EFFECTIVE APPLICATION OF PROJECT RISK MANAGEMENT

The Project Execution Plan (PEP) should also have a specific Risk Management sub-section dedicated to each of the defined sections of the PEP. This sub-section should define the risks associated with that component of the PEP and describe how the team will handle the defined risks. The sections of a "Best Practices" PEP Include:

Section 1 - PROJECT CHARTER

- Strategic and Business Objectives
- Project Objectives
- Scope of Work
- Support of Local Communities/ Workforce/ Enterprises

Section 2 - PROJECT PLANNING BASIS

- **Risk Management Plan**
- Organization Plan
- Master Plan/ Schedule
- Contracting Plan
- Funding and Cost Estimate Plan

Section 3 - PLAN FOR IMPLEMENTATION

- Design Management and Value Improving Practices(VIP) Plan
- Constructability Plan
- Procurement and Materials Management Plan
- Construction Management and Labor Plan
- Mechanical Completion/Commissioning Through Start Up Plan
- Regulatory and Environmental Management Plan
- Project Controls Plan
- Quality Management Plan
- Safety Management Plan
- Informational Management and Communication Plan
- Document Management Plan
- Operations Management Plan
- Plan to Capture Lessons Learned
- Special Factors Plan (Modularization, Project Suspension/ Restart Plan etc.)

EFFECTIVE APPLICATION OF PROJECT RISK MANAGEMENT

In a Best Practices PEP, the risk sub-section includes a write-up on the Risk Management Plan, the identified risk, what impact the risk will have on the project (+/-), how the team intends to address the risk, who is responsible for implementing the mitigation approach, when the mitigation is intended to be completed and what benefits (quantifiable/tangible of cost/schedule improvements) should be expected as a result of implementing the recommended mitigation. This is often called the risk register and becomes part of the PEP, but is also updated. Risks are mitigated as other risks become apparent.

Risk Identification

As an initial step in the Risk process, the team needs to "know what it does not know". It needs to brainstorm project issues that can both negatively and positively impact the project and then determine how probable the risk occurrence will be, as well as the impact to the project if the risk occurs. This comes from experience. Having "been there, done that". It also requires research of Lessons Learned from previous projects (inside and outside of the company) which will, at the very least, stimulate thought provoking issues that the team can evaluate for possible applications to their project(s).

So, based on the above discussion of limited experienced resources across the industry, how do we know what we do not know? How do organizations today share the limited project experience across projects to gain the knowledge and input for all projects rather than targeting just a few projects?

A traditional step includes implementation of a Risk Identification workshop in coordination with the Project Execution Planning workshop. As the team defines the execution approach that best suits the project and business objectives, various risks would be unfurled and documented as part of the PEP. The risks would then be assigned to responsible parties. These parties, together with the project team, would define mitigation strategies for each of the identified risks. This approach works well when the project team is made up of highly experienced resources representing their respective disciplines. Today's teams do not necessarily have this cross section of experience and must rely on other risk identification techniques.

Today, we rely heavily on checklists of questions and Lessons Learned related to various aspects of the project scope and execution strategy. These checklists only go so far – as the individual asking the questions may not see the inherent risk in how these questions are answered.

There is one technique that has proven effective but takes a mature organization to assure appropriate implementation. What some organizations have realized is that, with the limited pool of experienced project management resources available, an effective approach to gaining the best use of their experience is not necessarily to put them on a single project, but to have them act as an independent review resource that all projects must go through at key junctures in the project. The experienced individuals are asked to "Put their black hats on" and see what issues they see in the scope definition, execution plan, etc. of the project as it moves through the companies. These individuals need to have the intestinal fortitude to drill down into details at key junctures in the project to better prepare the project team for the future project phases. This function is often part of a corporate-wide Project Management Organization (PMO), which is a shared service across the organization.

As projects become more internationally integrated and technologically complex, some of the consistent risks identified in today's project environment include:

- Limited experienced planning, oversight and execution resources
- Competing business demands
- Significant competition for limited pool of qualified Engineering and Construction contractor resources
- Inadequate scope and/or site condition definition
- Escalation variability
- Supplier/Fabricator shop loading (over-commitment impacting delivery schedules)
- Labor Productivity – dwindling supply of qualified craft workers
- Extreme sensitivity to economic conditions/ capital investment reticence
- Discipline to follow established strategies, plans, processes and procedures (schedule pressures)
- Ill-defined contractual requirements (scope of work and services)
- Insufficient recognition of host country requirements/project environment (work rules, labor availability/qualifications, etc.)
- Logistical challenges (world-wide procurement, customs/duties, transportation, infrastructure at location, etc.)
- Learning curve costs (just-in-time recruiting, on-boarding, orientation)
- Staff turnover
- "First time" consortium/"partnership"/"alliance" arrangements – integration, coordination, systems and procedure compatibility
- Others...

These are but a few examples of the risks and challenges facing business and project ventures, particularly in mega-projects and/or international/multi-national situations/endeavors.

EFFECTIVE APPLICATION OF PROJECT RISK MANAGEMENT

Risks today that are typically overlooked are often related to the execution environment more so than with the technology and processes involved. As stated previously, lack of experienced project management resources is a huge risk issue today. Availability of competent international suppliers that can deliver the quality and timeliness required is another major concern, as is availability of contractor teams that have worked together in the past and are not thrown together solely for the project at hand. Multi-cultural execution teams also represent some unique challenges (in some cases).

Qualitative Risk Analysis

Identifying project risk is only part of the battle; it is time and cost prohibitive to mitigate all possible project risks. With this in mind, the team must prioritize project risks and select the most effective Risk Response for each risk based on business and project goals and objectives. The prioritization process must include a measurement of the impact the risk will have, if it occurs, as well as the probability that the risk will occur. A typical risk matrix is included below as Figure 4:

Figure 4

Risk Matrix

Scope of use: Utilized to assess, evaluate and categorize risks associated with capital project planning and execution, and Major Projects Contracts, Purchase Orders.

Project Specific

Consequence
(Note: For purpose of selecting impact of event, utilize most severe outcome of Project Specific and Corporate consequences analyzed)

Likelihood
How often would it be expected to have similar causes or circumstances aligned for the event with the defined consequences to re-occur? Please make reference to site, company and industry historical data to help predict the re-occurrence frequency of such an event.

Project Specific

Consequences

				Event will almost surely not happen	Event will quite unlikely happen	Event will quite likely happen	Event will very likely happen	Event will almost surely happen
Cost/Financial Impact	Schedule	Scope	Severity	**Very Low** --- < 5%	**Low** --- =>5% and < 20%	**Medium** =>20% & < 50%	**High** --- =>50% & < 80%	**Very High** --- =>80%
Cost Δ > $10MM or Cost Δ > 5%*Capex; which ever is smaller	> 6 months	Scope or Quality does not meet performance or business expectations	Very High	Moderate	Significant	Significant	Critical	Critical
1MM <Cost Δ ≤10MM; or 2%*Capex< Cost Δ ≤ 5%*Capex; which ever is smaller	> 3 months but ≤ 6 months	Scope or Quality unacceptable to Project Manager	High	Negligible	Moderate	Significant	Significant	Critical
100K <Cost Δ ≤1MM; or 1%*Capex< Cost Δ ≤ 2%*Capex; which ever is smaller	> 1 month but ≤3 months	Significant impact on scope or quality of major project deliverable and/or component	Medium	Negligible	Moderate	Moderate	Significant	Significant
10K <Cost Δ ≤100K; or 0.5%*Capex< Cost Δ ≤ 1%*Capex; which ever is smaller	> 1 Week but ≤ 1 month	Impact on scope or quality of minor project deliverable and/or component	Low	Negligible	Negligible	Moderate	Moderate	Significant
0 <Cost Δ ≤10K; or 0%*Capex< Cost Δ ≤ 0.5%*Capex; which ever is smaller	Negligible	No noticeable scope or quality impact	Very Low	Negligible	Negligible	Negligible	Negligible	Moderate

This too is a function where the above defined shared PMO risk function can be of benefit. The experienced project representative can bring some realism to the quantitative assessment of the identified risks. Inexperience can lead to downplaying what might seem to be minor issues, which if thoroughly analyzed, could have major impact to project success potential.

Quantitative Risk Analysis

The team must also measure the impact a risk might have on the project cost and schedule. The main issue here is that one risk may impact a number of project activities. As an example, site logistics or poor transportation to and from the site which might include limited road accessibility, multiple bridges and overpasses, poor road conditions, etc., will not only impact delivery of equipment and materials but also may cause construction productivity issues (people cannot get to the site in a timely manner), emergency response timing, local pubic relations issues, etc. These impacts all need to be taken into consideration when quantitatively measuring the impact of an identified risk.

There are a number of excellent risk analysis tools on the market that, when used properly, can assist the project team in measuring the impact risks have on the project cost and schedule. We will not go into their functionality in this paper, but many use a Monte Carlo simulation approach to develop risk profiles for both cost and schedule. This type of analysis assists the team in selecting which Risk Response effort provides the most project/business benefit for the investment. The project team needs to be well-versed in utilizing these tools, as they can lead to a misunderstanding of the impacts of risk if applied incorrectly. Owner management also needs to define what level of cost and schedule probability the project must "shoot for", as this will affect how much time and cost contingency are applied to the project.

If a selected risk mitigation strategy is applied, the cost, time and resources required to implement this strategy must also be considered, quantified and included in the base cost estimate and schedule. This is often overlooked and leads to false risk analysis readings.

Risk Response Planning

Probably the most important aspect of Risk Management is what the team will do to address the risk. Identification and measurement are critical but the real test is in how the team addresses the risks. There are a number of risk response options which are outlined in Figure 5 below:

Figure 5

Responding to Risk

RESPONSE TYPE	EXAMPLES
Avoidance	Choosing only high probability options, eliminating high risk approaches on parallel programs, screening subcontractors.
Prevention	Focusing on preventive measures, extensive potential problem analyses, extensive bench testing, mockup, and simulation work
Reduction of Impact /Mitigation	Focus on contingency planning, large contingency funding, frequent progress reviews
Compromising	Tending toward low risk options, choosing low return options, pursuing redundant paths
Transfer	Through subcontracting, warranties, performance bonds, insurance
Assumption	Any decision to proceed implies assumption of risk

Often risk registers are developed but not implemented; risk responses are identified, but too late. Many times the symptom is addressed but not the root cause (example: poor productivity in the field - the owner authorizes more resources but the true causes were engineering quality and/or late material deliveries).

In the case of engineering quality, the risk may be avoided by the selection of high caliber technical personnel within contractual terms.

Risk Monitoring and Control

Once the Risk Response approach has been implemented, management and project team representatives need to be assured that the selected response is doing what it intended to do. Project Control plays a key role here. The team needs to be assured that the money, resources and time allocated to a risk mitigation option are successfully achieving the benefits expected. If not, the team needs to reevaluate the option and develop an alternate approach.

One of the main issues we have today is the weakness in project controls across the industry. Of all the 'brain drain' suffered across the project disciplines, the degradation of the project control capability has been the greatest. This is a risk in itself. The team needs to be assured that it has a strong in-house, project control capability to assure the team can measure the effectiveness of the mitigations options.

EFFECTIVE APPLICATION OF PROJECT RISK MANAGEMENT

Why Is Risk Management so Important Today?

Today's projects are more complex, involve more planning and execution entities, are executed in a more challenging environment and involve multi-national resources. These issues, combined with a less experienced workforce are a recipe for disaster. Identification, measurement and response planning can aid in eliminating and/or minimizing the negative impact that risks may have on project success.

Who is responsible?

Risks are shared by the owner, contractors, suppliers, and other project stakeholders. Ultimately, the owner bears the burden of accountability for all project risks. At the end of the day, it's the owner's money. Product delivery is an owner issue (especially if they have made sales agreements and commitments to their customers). Quality becomes an owner issue – as the owner has to operate the facility for the next 20 – 30 years.

In Conclusion

A Project Manager in today's world cannot effectively manage a project without also managing the project risks. How aggressively one tackles the risk is dependent on the business/project drivers for the project, scope definition, execution team experience, etc., and is all part of the project execution strategy defined in the planning process. Risk management is one of the most critical skill sets a Project Manager needs to successfully manage projects in today's environment. Mastering this knowledge area goes a long way toward assuring achievement of project goals, objectives and success factor requirements.

(308f) Detection of Biomarkers for Different Diseases On Biosensor Surfaces

Ajit Sadana[1], Lokesh Taneja[2] and Kennon Shelton[2], (1)Chemical Engineering, University of Mississippi, University, MS, (2)University of Mississippi, University, MS

A fractal analysis is used to analyse the binding and dissociation (if applicable) kinetics of biomarkers to different biosensor surfaces. The dual-fractal analysis is used only if the single-fractal analysis does not provide an adequate fit.

For the binding and dissociation of IFN-gamma in solution to the aptamer modification (Tuleouva et al,. 2010), and for a single-fractal analysis, the (a) binding rate coefficient, k exhibits close to a third (equal to 3.01) order of dependence on the fractal dimension or the degree of heterogeneity that exists on the biosensor surface, and (b) the affinity, K (k/k_d) exhibits close to a fifth (equal to 5.064) order of dependence on the ratio of fractal dimensions, D_f/D_{fd}. This indicates that both the binding rate coefficient, k and the affinity, K are very sensitive to the nature or the degree of heterogeneity that exists on the biosensor surface.

For the binding of different concentrations of CA-II anhydrase in solution the binding of rate coefficient, k exhibits a mild (equal to 0.435) order of dependence on the CA-II anhydrase concentration in solution, (a) the binding rate coefficient, k exhibits close to a second (equal to 2.011) order of dependence on the fractal dimension, D_f that exists on the biosensor surface, and (b) the ratio of the binding rate coefficients, k_2/k_1 exhibits higher than a third (equal to 3.386) order of dependence on the ratio of fractal dimensions, D_{f2}/D_{f1}.

The above relationships presented are typical of the ones presented for the biomarkers for the other diseases analysed and for this presentation. They provided a means by which these rate coefficients or affinities may be manipulated in desired directions in order to improve the different biosensor performance parameters.

See more of this Session: Micro and Nanofabricated Sensors II
See more of this Group/Topical: Topical 9: Sensors

Who Are You Leading? How Do You Know?

Jack Hipple

Innovation-TRIZ
Tampa, FL

Successful leadership requires a clear vision, committed followers, and resources. Without all three the leadership goal may not be accomplished. Very frequently, a clear vision is established and resources are provided but the followers do not follow due to a disconnect between the style of the leader and the styles of the followers and team members. This situation can be resolved by changing the followers, but this can be costly and time consuming. It is not always possible for a leader to choose or change the followers that are needed. Command and control is one approach and the success of this can be a function of the supply and demand of the job market at the time. A far more reliable and robust approach is for the leader to take the time to understand his or her own style and behavior preference as well as that of their followers. Understanding these can help a leader be successful under any circumstance with any group. This presentation will overview some key behavior preference and style measurements that accurately assess social and problem solving styles and how they can be used to assist leaders and followers in achieving this goal.

Leadership and People

We use this term all the time without giving sufficient thought to what it really means. It may mean getting a group of people to see something that is not obvious to them, but is important to an organization and its senior management. It may be responsibility for leading a multi-national and multi-cultural team in building a multi-billion dollar capital project involving thousands of engineers in many locations. It may mean trying to figure out what the next new business, product, or service an organization should invest in when there is no clear indication of what it should be. However, all these responsibilities, and many other leadership roles involve one thing in common—the influencing and direction of people. A leader frequently does not get to choose who is on their team and occasionally needs the support of people who do not directly report to the leader. People are different. No two individuals are the same. They have technical backgrounds, different professional experiences, different personal preferences, and different biases. In order to lead a small team or a large multi-national group, it is imperative that a leader understand these differences for two primary reasons:

1. Task assignment. Everyone is not good at everything. Everyone has different preferences for what they would prefer to do. Assigning people responsibilities to people who do not enjoy that type of work when not necessary will only produce a drop in productivity and team morale.
2. Team relations. The time that a team spends together is valuable and expensive. If people on a team and its leader do not understand the differences in styles and

preferences, a certain percentage of the time and energy naturally present within each individual and the team will be wasted in arguments and debates relating to perspective, objectives, resources, etc. There will never be an answer that everyone will agree to, but an understanding of why a particular individual feels the way they do can go a long way to making everyone feel that their contribution and perspective is appreciated.

Team Member Preferences and Styles

Over the years, many different individual assessment instruments have been developed which are useful for assessing differences in social and problem solving styles. Both are important in a typical engineering team as the leader would like the team to "get along" but also solve a problem or design a process, product, or facility. These are different things. Social style relates to how individuals relate to each other in a social context, not necessarily one that includes technical problems. Problem solving style relates to how individuals analyze problems and how they solve them, independent of who they may be working with who is also involved in the problem solving or design challenge. It is possible for individuals to have identical social styles as measured by an assessment such as the Myers Briggs assessment and to have very different problem solving styles as measured by an assessment instrument such as the Kirton KAI™

Social Styles

The Myers Briggs MBTI™ was the original social style assessment instrument and has been given to millions of individuals. It has successors such as Insights™ and 16Types™ which provide feedback in the same summary form. Regardless of which instrument is used, a person's social style can be characterized along the following continuum:

Extroversion/Introversion (E vs. I) This describes the outward vs. inward focus of an individual. An extrovert enjoys socialization, verbal communication, cold calling, possibly parties, and other occasions and events where outward communication is enjoyed and expected. A strong introvert, on the other hand, may prefer to read a book, surf the web for information and become very nervous in social situations where direct oral communication seems to be required.

Sensor/iNtuitor (S vs. "N").Note: An "N" is used here, rather than an "I" to avoid confusion with the "I" used in the previous category. A Sensor is strongly concerned with data, facts, and measurable aspects of a problem or situation whereas an Intuitor is equally concerned with aspects of a situation that may not be directly measurable such as implications, indirect effects, gut feelings, emotions, and undefined possibilities that cannot be immediately quantified. This characteristic primarily applies to how information and data is taken into a mental analysis as opposed to what comes out afterwards. A simple question to ask someone to find out which one of these they may be is to ask them to describe a leaf. A sensor will say things like “green” or “scaly”, terms that are physically descriptive. An intuitor will describe it as “light” or “reflective”.

Thinking/Feeling (T vs. F) This attribute characterizes how someone treats the analysis of the information received. A “T” person will be much more concerned with the impact of a decision on economics, structure, and costs while an “F” person will be just as strongly concerned about changes might affect the personal lives of individuals and their careers. In a downsizing situation, the “T” will want to know how much money will be saved and what the outplacement services will cost, while the “F” will want to know about the impact on peoples’ families and their psychological well being.

Judging/Perceiving (J vs. P) This characteristic describes the focus of an individual and their approach to implementation and planning. A “J” will be very concerned with schedules, commitments, deadlines, and critical path diagrams. A “P” will be much more open to loose schedules, leaving the door open for changes and more comfortable with uncertainty. Deadlines are critical to the “J” and far less important to the “P” who is more worried about missing possibilities that cannot be seen as opposed to meeting commitments to do what is known.

You can imagine, as a leader, the challenges in managing a team where these characteristics were not uniform. It’s easy to work with and manage people that are like us, but that does not necessarily make for the best team structure. If everyone thinks alike, there is no one to ask the difficult questions or propose alternatives and options

The social style assessment will come back, independent of the type of instrument used, as four letters, such as ESTJ or INTP. Though the descriptions above emphasize the extremes, it is possible for an individual to be borderline or in the middle. There can be small changes in an individual’s preference over time, but not large ones. This type of assessment also provides an insight as to how a person will respond under stressful situations. For example, a strong introvert will become very extroverted when fundamental beliefs or values are questioned. Obviously, this type of situation can occur within a project team and if social style preferences are known and understood, these types of situations can be handled more effectively.

Problem Solving Style

An individual’s problem solving style can be measured very accurately with an assessment instrument known as the Kirton KAI™. This assessment, consisting of 32 simple questions, analyzes a person’s approach to problem solving. The assessment comes back as a number from 32-160 with an average of around 90 and a 2 sigma deviation between 70-110. A lower KAI™ number would describe someone who problem solves with known parameters, likes schedules and agendas, and generates “filtered” ideas whose probability of being practical is high. They are described as an “adapter”. At the other extreme would be someone who tends to generate a larger quantity of ideas without necessarily worrying about their practicality, is less concerned about agendas and schedules, and will "bi-associate" ideas and make connections between disparate concepts not made by others. In KAI™ jargon, this person is described as an innovator. A simple question to ask someone to get a sense of where they are on the KAI™ scale is what they would do as a tea bag manufacturing plant manager who has just lost a major customer amounting to over 2/3 of the plant's business. An adaptor will suggest ideas relating to cost reduction and layoffs and encouraging sales and marketing to find some more tea bad customers. An "innovator" will ask what new product ideas might come from packaging something else in the bad or what else the bag fabric could be used for.

As mentioned earlier, people with the same social style profile can have differing problem solving style profiles. This can set up a situation where a team leader thinks that everyone on a team "likes" each other and their job will be easy, but if the team's challenges include serious problem solving, things may not be as easy as anticipated.

Leadership Situations

Let's consider some situations and the implications of not understanding the nature of the team being "led". First, an international new product introduction. If the leader's future income is dependent upon how fast and how many countries a new product can be introduced into, there is a high level of risk, especially if manufacturing and distribution are under contract in some sense. If the team members do not have the same stomach for risk and "fuzzy" situations, they may throw up artificial roadblocks and record "remember I told you" notes to be vindicated later if the introduction does not go as planned. If the leader knows about this style preference ahead of time, then a rational discussion can be had regarding how the risks can be minimized.

Consider a multi-country engineering team whose meetings may be as often over the web as in person. These various social and problem solving types are not uniformly distributed and can be affected by cultural biases. It is critical, especially when deadlines and commitments are involved, to make sure that a leader ensures that the meaning of deadlines is clear. A "J" will understand easily while a "P" may interpret a deadline as simply a target or a goal. A high scoring KAI person may take risks that seem acceptable to him/her but may not be acceptable to the team leader. Interpretation of boundaries will vary and the language needs to be understood at the beginning of a project, not in the middle of a crisis.

To summarize, it is critical that a team leader understand both social and problem solving styles of the people on their teams. Without such an understanding, there is great risk of misunderstood goals, decision making based on misunderstood criteria, and problem solving done with either too little or too large a boundary condition.

Ethical Project Management

Dr. Eldon R. Larsen
Marshall University

ABSTRACT

Project managers are often faced with situations that present ethical dilemmas. These situations can be anything from very simple, like something in a conversation, to something very complex, like how to accept or not accept a gift. This paper presents some principles and suggestions regarding ethics in project management. It also highlights the main concepts in the Project Management Institute's ® Code of Ethics and Professional Conduct.

INTRODUCTION

Many people have written about ethics. The objective of this paper is not to re-hash past studies. Rather, I attempt to present significant concepts and principles through both quotations and discussion that will help the reader gain a vision of what it means to be ethical in one's behavior. This applies very much to project managers, as they are frequently making decisions that have ethical questions surrounding them.

CARDINAL RULES OF ETHICAL BEHAVIOR

Following are a few quotations in which are embedded basic principles and rules of ethical behavior. For several of the quotations are given "sub-quotations" that are closely related, but are expressed in entirely different ways.

- "Therefore all things whatsoever ye would that men should do to you, do ye even so to them" (Matthew 7:12)
    - "You can easily judge the character of a man by how he treats those who can do nothing for him." (James D. Miles)
- "And above all things have fervent charity among yourselves: for charity shall cover the multitude of sins." (I Peter 4:8)
    - "When I do good, I feel good. When I do bad, I feel bad. That's my religion." (Abraham Lincoln)
- "Thou shalt love thy neighbor as thyself." (Leviticus 19:18)
    - "Laws control the lesser man. Right controls the greater one." (Chinese proverb)
- "To thine own self be true, and it must follow, as the night the day, thou canst not then be false to any man." (William Shakespeare, *Hamlet)*

- "To know what is right and not do it is the worst cowardice." (Confucius)

FOOTNOTES TO THE CARDINAL RULES

- Though you legally can do something to someone else, it might not be alright to do it to him.
- Ethical people normally follow the laws, but are not constrained from going beyond the bare minimum of the law to do what is right
- Good ethical behavior is at the heart of excellent leadership—you do what is right because it is right, not because it benefits you monetarily or otherwise
    - "Keep true, never be ashamed of doing right; decide on what you think is right and stick to it." (*George Eliot)*
- "If moral behavior were simply following rules, we could program a computer to be moral." (Samuel P. Ginder)

ETHICAL PROJECT MANAGEMENT AND PRINCIPLES

Ethical project management involves right behavior personally, legally, socially, and environmentally. Inconsistencies in one area can affect behavior in another area. Consistency of ethical behavior in all areas strengthens a person's integrity.

We normally consider the goal of project management to be to achieve project objectives. Some questions we should ask ourselves, however, are the following:

- How well do we meet the objectives, and within what constraints?
- Does it matter how we get there?
- Are some objectives more important than others?
- How do we treat team members? Stakeholders?

How might ethical considerations affect the answers to the above questions?

Many organizations have a code of ethics by which they expect their participants to abide. However, it is important to consider how the code was developed. I suggest that you should always questions a code of ethics which is based solely upon the premise that right or wrong is determined either by opinion or by statute. Truth is true whether people agree with it or not. And even making a statute does not change that truth is still true, and false is false. True principles will always be true, no matter how hard we might try to ignore them, change them, or go against them.

The fact that true principles will always be true can present us with a *principles dilemma*. One of the challenges we face is when principles seem to conflict with each other. For example, is it right to tell a lie in order to save an innocent person's life? Many people would probably say "Yes." This leads us to realize that one principle might be more important than another, even though they are both true (Don't kill. Don't tell a lie.). It is often our conscience and our experience that tell us what to do in these circumstances. In fact, good ethical behavior is closely tied to a sensitive conscience.

PRINCIPLES AND VALUES

Principles and values are not exactly the same thing, although they can be very closely related. Most values are built upon some principles. Values are like roadmaps that guide us in our decision-making processes. Therefore, it is important for us to identify and understand our values—companies and organizations need to understand their own core values. Once we better understand our values, we can more easily make an ethical decision through some kind of structured analysis.

Cultural norms can affect what is considered acceptable or not. While principles (truth) will always be true, the application of the principles might change significantly, depending upon the situation. Values (how we perceive life) sometimes vary with culture, religion, upbringing, etc., and therefore change the situation. They relate to perspective. So, it is important to understand cultures when considering what decision we face is ethical. We don't have to break principles when we adjust for different values—they are not the same thing.

Project management ethical considerations often involve the following questions.

- Are you obeying the law? Are you following guidelines provided by your organization and the customer's organization? Which one should take priority?
- Are you treating other people the way you would want to be treated, and would you do it either in private or in public and still have personal peace?
- What is driving your actions? What if everyone did what you are considering doing? Would that be good?
- Is your choice the right thing to do, according to the principles you know to be true?

There are also some legal considerations when dealing with ethical questions. For example:

- "The Foreign Corrupt Practices Act – FCPA, 1977, states with respect to bribery that it is a crime for a bribe to be made to a foreign official or political party for the purpose of obtaining or retaining business or for directing business to another person even if this flows through an intermediary or consultant.
- "However, not all payments are defined as "bribes." Payments for "routine" government action are allowed under the FCPA including the dispensing of permits, licenses and police protection as noted in the example."[1]

Following is an ethical, legal situation to illustrate some of the challenges in making ethical decisions:

- Country A allows companies to perform drug testing on employees, but allow employees to refuse to be tested
- In Country B, companies can mandate that employees and contractors submit to drug testing, and refusal can be grounds for dismissal
- A Country A contractor is asked to perform work for a company in Country B, that asks the contractor to submit to a drug test. He refuses. Right choice or not?

Even though the contractor has the right to refuse drug testing according to the policies of his own home country A, because he is doing work for company in country B he should follow the more stringent policies of country B.

The Project Management Institute's Code of Ethics and Professional Conduct

The Project Management Institute's Code of Ethics and Professional Conduct[2] includes many fine points of ethical behavior for project managers. For this reason, it is useful to summarize them here.

1. Take responsibility for the decisions we make or don't make, and for the resulting consequences
2. When making decisions, take into account the best interests of society, public safety, and the environment
3. Keep commitments
4. If we make a mistake, take ownership and responsibility, and correct it. If others make mistakes, communicate appropriately as quickly as possible
5. Protect proprietary or confidential information
6. Be informed regarding policies, laws, regulations related to our work, profession, and volunteer activities
7. When substantiated by facts, report unethical or illegal conduct to appropriate management and, if necessary, to those affected by the conduct
8. Be informed about the customs of others and avoid being disrespectful
9. Listen to other people's opinions and views and try to understand them
10. When a conflict occurs, deal directly with the people involved
11. Be professional, even if this conduct is not reciprocated
12. Negotiate in good faith
13. Do not use power to benefit personally at the expense of others
14. Do not act abusively
15. Respect the rights of others

16. Demonstrate transparency in decision making
17. Be impartial as much as possible
18. Allow equal access to all who are authorized to have information
19. Make opportunities equally available to qualified candidates
20. Actively and fully disclose potential conflicts of interest to appropriate stakeholders
21. Refrain from engaging in decisions or influencing others when we have a conflict of interest, unless or until: we have made full disclosure to affected parties; we have an approved mitigation plan; and we have obtained consent from affected stakeholders to proceed
22. Do not hire/fire, reward/punish, or award/deny contracts based on personal considerations like favoritism, nepotism, bribery, etc.
23. Do not discriminate against other people based on things they cannot control or based on their religion
24. Apply your organization's rules without favoritism or prejudice
25. Seek the truth
26. Communicate truthfully
27. Provide timely, accurate information
28. Make implied or explicit commitments in good faith
29. Strive to create a non-threatening environment for people to tell the truth

SUMMARY

Ethical project management involves many facets of application, all of which should be based upon foundational principles of good management and human relations. The basic rules or principles of ethical behavior have existed for thousands of years and will continue to exist for thousands more. Truly successful project management applies these ethical principles fairly according to the situation and the needs of the people involved.

REFERENCES

1. http://www.bia.ca/articles/pj-project-managers-ethical-dilemma.htm
2. http://www.pmi.org

Effective Quantitative Decision-Making based on Lowering Cost of Capital and ROI thru Safety Eng Studies

Gutiérrez Capriles Glenda J
***Risk Project Analysis Leader Engineer Advisor**
Avenida Circunvalación del Sol, Urbanización Santa Paula, Torre Inelectra, piso 7, Municipio Baruta, Estado Miranda, Venezuela
glenda.gutierrez@inelectra.com

Prepared for Presentation at
AIChE 2011 Annual Meeting
Minneapolis, MN, United States
October 16-21, 2011

UNPUBLISHED

Effective Quantitative Decision-Making based on Lowering Cost of Capital and ROI thru Safety Eng Studies

Gutiérrez Capriles Glenda J
***Risk Project Analysis Leader Engineer Advisor**
Avenida Circunvalación del Sol, Urbanización Santa Paula, Torre Inelectra, piso 7, Municipio Baruta, Estado Miranda, Venezuela
glenda.gutierrez@inelectra.com

Keywords: Management of projects, processes of management, areas of management, phases or service life of the project, Decision-making methodology, instrumented protective systems (IPS), spills, return on investment (ROI), hazardous events, fire zones

Abstract

The intention of this article is to reflect on the main aspects that affect the performance of the projects, for such effect took place a diagnosis of the schemes of management used by the company of the consulting sector. For the study one approached the areas of generation business, petroleum, refining, petrochemical, technology, etc of inelectra and the operational area of a production facility.

The process industry has adopted automation to improve product quality and production rates, to reduce the potential for operator error, and to decrease resource requirements. Process automation includes many different instrumented systems, such as process controls, alarms, instrumented protective systems (IPS), and emergency shutdown systems. These systems range from simple hardwired systems to complex programmable electronic systems.

Good design leads to safe operation by anticipating potential errors and failures and implementing means to detect their presence, so that they can be corrected once implemented, the increased complexity may require a significantly greater level of attention during assessment, design, inspection, testing, and maintenance.. Good design ensures that the core attributes of independence, functionality, integrity, auditability, reliability, access security, and management of change are addressed consistent with owner/operator requirements. Poorly implemented design can lead to significant hazardous events, involving impact to people, the environment, and plant assets.

Various valuable information scattered all over CPI (Chemical Process Industry) needs to be integrated and retrieved efficiently to manage technique, facility and personnel. This paper focuses on a new decision-making methodology of safety-related investment and management of main causes for accidents and failures. The methodology integrates safety technology, process data and business information in a petroleum and oil-refining industry. This methodology suggests a quantitative investment procedure to reduce safety accidents and plant failure cases within limited budget efficiently. Major factors are considered as accident history, human error, facility, service, operation, maintenance, emergency plan, and priority of investment, effectiveness of improvement, environmental concerns, loss control and cost.

1. Introduction

Any industry, principally oil-refining industry, have to keep safety management methods that ensure risk of fire, explosion and toxic material release which is very important in saving life, assets and environment. The right decision-making for improving safety in chemical plants requires more quantitative and systematic methods of analyzing previous accident records, process data and business information. This safety engineering studies are necessary to be assure to the whole decision-making procedure for project risk analysis evaluation, to prevent major accidents and to reduce the risk of a plant.

The initial design effort includes the prediction of potential hazards. Numerous types of hazard analysis methods have been proposed and used. Hazard analysis methods are classified by qualitatively and quantitatively. The qualitative methods include Check-lists, What-If Analysis, Failure Modes and Effects Analysis, Preliminary Hazards Analysis or HAZID and HAZOP. The quantitative methods include Event Tree Analysis, Fault Tree Analysis, Quantitative Risk Analysis, Reliability Analysis and Safety Integrity Level (SIL) / Safety Instrumented System (SIS).

New technology is of value only if it provides a suitable financial return. Many articles have been published on the potential economic benefits and return on investment (ROI) of facility and advanced automation technology in the process industries. Unfortunately the benefits claimed are often unrealistic and unsubstantiated; this leads to significant credibility issues when the forecast benefits are not achieved and to a lack of confidence on the part of management on proposals in this area. In this article a short review of the proper way to perform financial analysis as part of project risk analysis that was affected by improper safety engineering studies execution.

2. Review of Plant Economics

The first step is to examine the plant as a financial asset. We may think of plants as a collection of equipment and personnel that convert raw materials into products. From a financial point of view, a plant is an asset that consumes money and produces money, hopefully producing more than is consumed. The major monetary components are shown in the figure following:

Figure 1: The Plant as a Financial Asset

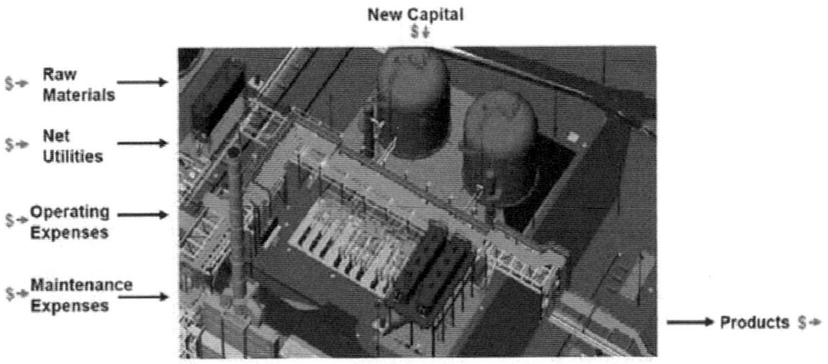

By convention, financial inputs are classified as either capital or expense. Expenses include all the ongoing costs of producing the products such as raw materials, net utilities (used less produced), operating costs, maintenance expenses, and other miscellaneous costs. Capital has two components – working and investment.
Investment capital refers to the cost of major equipment or system additions that will last for several years and can be depreciated for tax purposes. Working capital is the value of inventory and required net short term financial funding.

3. Return on Investment Capital taking into account also, Procurement and Safety

To properly prepare an investment evaluation we must understand the language and objectives of the plant's financial management. In this section a quick overview of the subject is given. For more background and a more comprehensive review.
Ultimately the management of a company has an obligation to increase the long term financial value of the corporation to its owners, who are normally the shareholders. This value can be measured in many ways but stock market valuation is one of the more important and ultimately is a function of financial performance. There are equally many ways of evaluating financial performance but the primary measure used here will be Return on Invested Capital (ROIC). This is not only a good internal measure but also correlates with long term stock market performance. The consulting company McKinsey (2003), recently published results of a study by of 130 publicly traded chemical companies in the US and Europe and their financial and market performance over 40 years. Their conclusion: "only returns on invested capital drive market-to-book valuations".

Figure 2: Return on Invested Capital

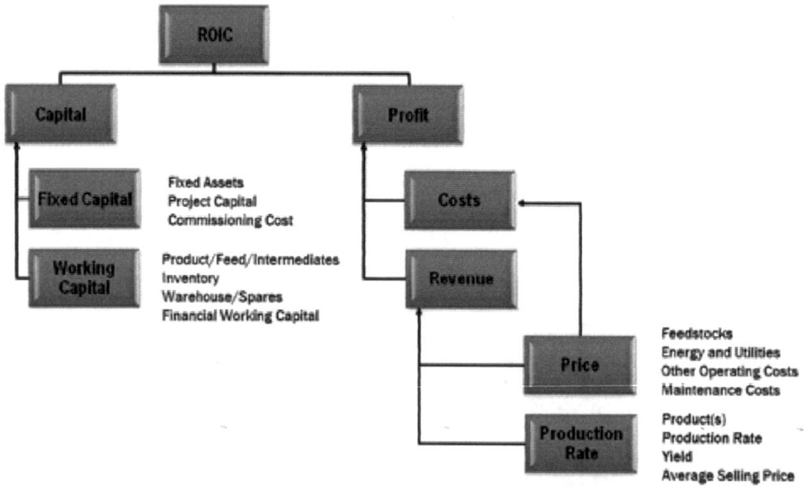

To increase ROIC, capital must be reduced or profit increased or preferably both at the same time. The primary areas where automation and advanced automation savings are normally found are illustrated in the boxes at the bottom of the figure. When a project is considered, all possible savings areas should be evaluated.

Potential capital savings include both fixed and working capital components. Potential project capital cost reductions due to automation choices can include savings in:

- Engineering
- Procurement Costs
- Purchase Price
- Installation, Configuration, Calibration & Commissioning
- Project Execution

Working capital can be reduced by reducing raw material, intermediates, and product inventories and also by reducing owned spares for equipment stocked in the warehouse. Capital deferred also results in savings due to the time value of money.

Deferred capital can result from longer equipment life due to better control or from more production from the same equipment which results in a postponed plant expansion.

Expenses can be reduced by lowering energy /utilities usage and decreasing raw material costs through increasing yields of desirable products. In special cases it may be possible to substitute a lower valued raw material.

There are several types of maintenance costs including:

- Scheduled Maintenance
- Unscheduled Call Out
- Major Shutdown

Reductions can occur in all types of maintenance costs due to improved automation performance and enhanced monitoring of process equipment. Specific savings can include reductions in:

- Unscheduled maintenance
- Number of routine checks
- Time to perform required maintenance tasks
- Maintenance materials purchase
- Number and costs of required tasks during scheduled shutdowns

Reduced off specification material can lower reprocessing costs. Demurrage is the cost charged by ships waiting to load. Waits beyond contracted amounts incur penalties. In selected cases it may also be possible to reduce average staffing levels.

Abnormal event prevention deserves some special comments. Reducing Health, Safety, and Environmental (HS&E) issues is at the top of every plant manger's agenda.

Improved automation is often a key to reducing these types of events. If your proposed investment will result in safer operation or reduced emissions it should definitely be included in the justification description even if no quantitative economic value can be attached to these improvements.

Next is the area of increased average selling price for products. Automation can contribute to this goal by increasing the yield of more valuable products. Reduced amounts of lower valued byproducts will increase average revenue per unit feed. Off specification material, which can be reduced through improved control, usually brings a discounted selling price. Occasionally an increase in product quality due to improved control will actually permit selling the existing product(s) at a higher price.

With regard to increased production, the key question is whether or not the additional product(s) can be sold. Increases are only financially valuable for production limited plants, i.e. those manufacturing products where the market can absorb the increases. If not, no benefits can be claimed for the improvements. Increased production can result from the capability, with better control, to operate closer to production limits at constant product quality. Increased production can also come from reduced unscheduled downtime for equipment due to better reliability, shorter batch cycle times, lower grade transition time, less product reblending, and decreases in scheduled shutdown duration and frequency. The shutdown frequency can be reduced, for example, by increasing a furnace run time between required decoking or not cleaning a heat exchanger prematurely.

To estimate the potential magnitude of the savings, examine the normal expenditures and look for specific quantifiable areas where savings are possible. Often historical data will provide a basis for this analysis. External consultants with experience in this area can also help insure that potential savings are not overlooked and that there is proper documentation for the estimated savings used for the justification.

3.1. The Cost of Tendering

Major equipment manufacturers have calculated that it costs them up to thousand dollars to bid on a complex tender. These costs are eventually passed on to the customer although be it indirectly.

Among the major contributing factors, are:

Long Lead-Time of the Process - There are many time consuming tasks that contribute to the long lead-time of tendering including: internal collaboration within a multi-disciplinary group of people within a single organization (but often geographically distributed), obtaining approvals (both financial and technical), preparing the tender specifications and documents, interfacing between customers and suppliers, interfacing with other companies, third party consultants, partners in a consortium and sub-contractors, assembling accurate bidding documents and calculating prices and time estimations.

The Quality of Tendering and Bidding Documents - This is a crucial element for successful and efficient tendering and involves criteria such as the clarity of requirements and specifications, compliance with quality certification standards, clarity of the bid, exhaustiveness, adequate structure, nice presentation, compliance with specific requirements set-out in the tender, completeness of the information provided, etc.

Inability to Quickly and Reliably Assess the Bids - Differences and inconsistencies in the bid formats make it very difficult to assess the bids and compare apples with apples. A large amount of time is often spent collating data from the various bidders into a format that allows for proper comparison and evaluation.

Reliability of the Cost and Time Estimate - Ultimately companies who are consistently accurate in their pricing and delivery estimates will win in the long run. The ability to provide reliable estimates depends both upon the capability of the bidders and the clarity and unambiguousness of the tenders. Clear, complete and non-ambiguous tenders will also naturally lead to lower bidding prices. As uncertainty diminishes so does the "safety price" or padding added to the actual estimate.

Monitoring Delivery of the Works - The ability of the customer to continuously and transparently monitor the current status of the works is a definite advantage in that it reduces delays and ensures delivery in compliance with the contract.

3.2 The ROI of Safety

Although it is difficult to prove that incidents will happen in the future, management and financial representatives of the company need to be aware of the consequences of an ineffective safety solution. By understanding the elements of risk management and the methodology of a safety investment analysis, safety engineers or risk management professionals can present the financial justification to help make operations safer and reduce both workplace injuries and insurance costs.

Integrating a safety solution to manage risk is a process that is best employed via up-front teamwork between operations and risk management, such as risk managers, safety professionals and claim analysts. In addition, it is important to understand risk and the different types of risk that exist on a plant floor. Risk awareness often leads to better decisions that can reduce soft costs. By providing an easy-to-be-safe environment for operators with safety solutions that are designed into the machine and focus on complete machine life-cycle requirements, companies are able to reduce indirect costs. In this paper, we will highlight key elements of effective risk management and the risks that affect the engineering, procurement, construction and operation of Production facility, discuss the implications on workers and provide ideas to measure potential savings that can be found in proactive safety investments.

Risk management is a two-step process that identifies and analyzes uncertainty in decision-making. It consists of determining risks that exist in an equipment investment and managing those risks in a way best-suited to achieve investment objectives. For managers and financial professionals investing in safety, proving the benefit of the safety investment is difficult since it could prevent injury and avoid costs for a company. Using the risk management process, professionals can determine the level of safety investment that meets their needs and would provide the best result for their situation.

An unplanned interruption can threaten a company's supply chain because of the decreasing amounts of raw materials readily available on the plant floor. These limited resources affect productivity levels and operation costs and threaten the overall health of a company as lower levels of inventory put manufacturers at increased risk for losing revenue each time a production line goes down. Effective safety programs can be implemented to help minimize this risk so manufacturers can better protect their supply chains.

Once the risks are fully defined and clearly understood, they must be designed out or mitigated to the greatest extent possible. Risk mitigation measures are necessary in order to physically improve the machine to prevent personal injury or property damage. Risk mitigation can be accomplished through a variety of activities, as demonstrated in Figure 2 below. One effective mean is to use safeguarding equipment such as installing physical guards or safeguarding control products, such as light curtains, safety relays and cable pull switches to reduce risk to employees.

Figure 3 – Hierarchy of Protective Measures

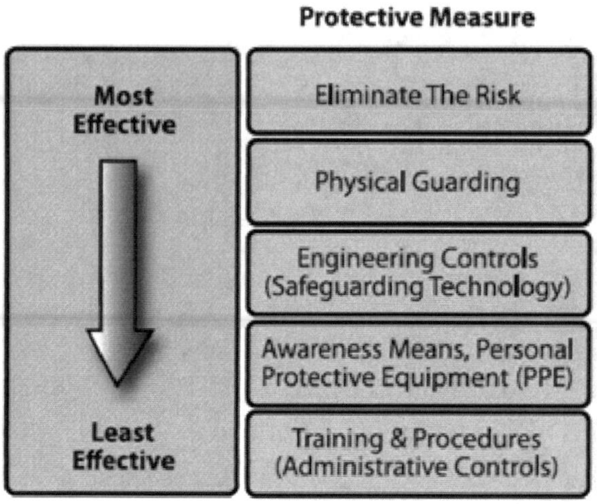

Using a formal risk assessment process also provides the benefit of documenting how the risks have been reduced or designed out via safeguarding solutions or other protective measures. The use and documentation of the risk assessment process is a critical tool for companies to show their due diligence and good engineering practices in providing a safe work environment. By effectively implementing and documenting the process, the resulting protective measures, and training programs, a company can potentially lower risks of litigation should an incident occur. After implementing and documenting the process, it is important to provide the appropriate training and supervision. It is crucial to that operators understand the safety measures, including usage of personal protective equipment. Operators must be trained.

Safety is not a luxury reserved for developed countries. One topic of discussion is this idea that safety and health programs benefit a company's bottom line. For many analysts, the bottom line is simple: "The CEOs 'who get it' recognize that a corporate commitment to safety has its costs, but, more important, it pays tremendous dividends in business and human capital."

So there is a clear connection between safety, quality, production, and profitability."

4. Investment Costs

Next investment costs must be estimated. When performing investment analysis, it is important to consider the full life cycle costs, not simply the initial purchase price, of new equipment and new software systems. Many studies have shown that more than two thirds of their life cycle costs occur after installation.

In the project initial, or capital costs area, include the following:

- Installation costs
- Required infrastructure upgrades
- Software licenses
- Application specific services – design, configuration, installation, integration and commissioning
- Project expenses – project management, purchasing
- Training – System support and end user

Also, ongoing maintenance and support cost estimates. Don't forget the last item. Make sure that you budget for ongoing training. The best long term job security today is maintaining you and your staff's skill levels and expertise current. It is easy to fall into the trap of being so busy with day to day problems that you lose sight of the longer term requirements.

5. How Can ROI be Useful?

Profits happen when a company operates effectively. We can tell that the management team is doing its job well if the company prospers, obtains funding, and rewards the suppliers of its funds.ROI is the principal tool used to evaluate how well (or poorly) management performs.

6. Safety Decision-Making Procedure

A framework for viewing management information systems is essential if an organization is to plan effectively and make sensible allocation of resources to information systems tasks. A framework is needed to provide a more efficient allocation of resource in safety information. This study suggests the procedure, development and information management of safety decision-making methodology in two different information areas as shown in Table 6.1. A framework is suggested for Decision-Making by using model structure and implementation procedure which differs sharply between the structured and no-structured information area.

Table 6.1. Framework of Safety Information Management System

	Safety	Process	Business
Structured	MSDS Regulations Standards Company Policy	P&ID PFD Plot Plan	Department Personal Facility Warehouse Buildings
Non-structured	Accidents Measures Risk Accident History Causes Analysis	Technical Reports Technical Audits Design Execution with Quality MOC History	Cost Strategy Planning Investment Priority Project Risk Analysis Benefit-Cost Analysis

CCPS (Center for Chemical Process Safety, 1998) in AIChE presents effective taxonomy of safety data for reliability analysis as shown in Figure 4.

The basic building blocks for the reliability analysis fall into two main classes. One is the data regarding specific events occurring in a piece of equipment. The other is the data describing the sample space of equipment. Event data in the first class is used as a numerator in the reliability calculation and inventory data in the latter class is used as the denominator.

Classification of safety accident is important while collecting safety information, so we classified it as accident results, direct and indirect causes of accident, primary causes, process status, utilities, departments etc.

Significance and operating variables also affect the business strategy in plant site also of Project Risk Analysis.

Figure 4. Database Taxonomy and Structure

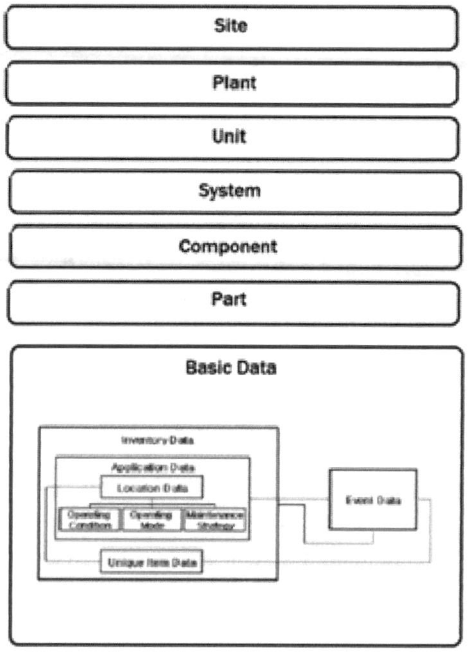

This study describes an overall systematic decision making procedure of preventing or minimizing safety accident under the given budget as shown in Figure 5. To decide the investment of safety related facility, two inputs are required: 1) the priority of investment and 2) the criteria of investment. To list up the priority of investment requires analyzing plant accident data and process data with the general safety information by using matrix computation method that we propose. To set up the criteria of investment requires surveying business-level executives regarding acceptable safety damage such as casualty, cost loss and business interruption.

Figure 5. Safety related Business Decision-Making Procedure.

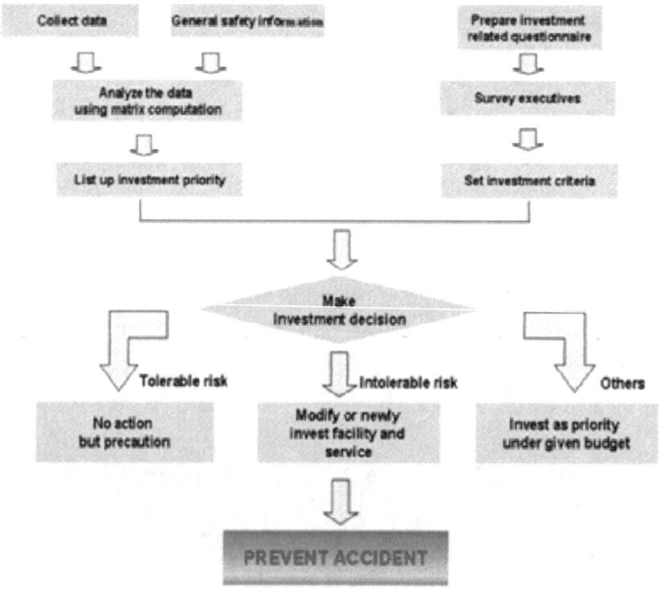

Based on the above two preparations such as the priority and the criteria, the decision is made according to the three risk probability regions: 1) definitely tolerable risk region, 2) definitely intolerable risk region and 3) other region in between the two. If the risk of a hazard is tolerable, the company does not need to invest any money into the relevant event and, in many cases, just careful attention is necessary. If the risk is definitely intolerable, the company must invest money to modify the process or purchase new facility and related service regardless of the budget if the plant needs to be run continuously. If the risk is between definitely tolerable and intolerable region, the investment is negotiable with the given budget that is, the investment to each event is selected by the priority under the condition that the given budget allows. As following this decision procedure, the company invests money only to necessary facility and service for preventing safety accident under the given budget.

7. Study Case: Surface Upstream Project

7.1 Scope

The whole Station Upgrade Project comprises the installation of new facilities, and revamp of existing ones needed to increase the produced water treatment capacity during production of Extra Heavy Oil (EHO) of 100 KSBPD to 200 KSBPD.
In more detail (Figure 6), MSUP includes:

- A new Diluted Crude Oil (DCO) Treating Train (Train C), capable of treating 95 KSBPD of EHO diluted with 34.4 KSBPD of Diluent, with 77 KSBPD of associated produced water.
- A new Produced Water Treating Plant of 200 KSBPD.
- Utilities revamp, including a new 25 MW Turbogenerator and 100 KBBL Diluent tank.
- Minor upgrades in existing DCO Treating Trains.
- New Flare
- New Fuel Gas Skid
- New Chemical Injection Packages
- New Desanding Systems
- New Open and Closed Drain System
- New Control Room (MSCR)

7.2 Location

Main Station facilities are located to 7,5 km South-Eastern from San Diego de Cabrutica, near from region Pariaguán, to South Anzoátegui State, in Venezuela.

7.3 Facilities Description

Main Station facilities are part of the Upstream Section of an overall heavy oil production and upgrading concept that will lead to the production of Synthetic Crude. Development of this concept is under way through two different Engineering Projects. The project that deals with the production and initial conditioning of the heavy oil is referred to as the Upstream Surface

Facilities Project, while the one that deals with the transformation or upgrading of the heavy oil to Synthetic Crude is referred to as the Downstream Facilities Project.
The major elements of the Upstream Section are:

- Oil Wells, grouped in Clusters, to produce the heavy crude.
- Diluent System, to dilute the heavy crude at the well head and improve its pumpability.
- Gathering Network, to collect and send the diluted heavy crude for initial conditioning.
- Main Station, to carry out the initial conditioning.

The facilities in the Main Station Upgrade Project are divided into Third Oil Dehydration Train called "Train C", a Production Water Treatment Plant (PWTP), and some utilities revamp (fuel gas, Diluent, open and closed drains).
The existing MS Oil Processing Trains (A and B) are not capable to handle the new production target 246 KSBPD EHO, 89 KSBPD Diluent and 200 KSBPD Water due to the increasing water cut. In order to handle the fluids coming from clusters a Third train is required.

- Export Pipeline, to send the conditioned heavy crude to an intermediate Pump Station, operated by another party that will finally get the oil to the Upgrader Section.

Main Station's primary objective is to process the incoming fluids from the gathering network, i.e. oil, associated gas and water, so that oil can be consistently exported to the Upgrader within the required specs and at proper pumping conditions.
Oil to be exported from Main Station shall have a maximum water cut of 2%, a Reid Vapor Pressure no greater than 0.76 bar_a (11 psi_a), a specific gravity of 0.953 (17° API) and shall be pumped at approximately 89 °C and 25 barg.
The crude oil is separated from the gas and is stabilized to less than 0.76 bar_a (11 psi_a) RVP and the water content is reduced to less than 2% vol. The API gravity of the crude oil is adjusted by adding diluents to 17 °API and is then pumped to the export pipeline. The facilities at the Main Station are continuously manned.
The fluids coming from the gathering network enter Main Station and split into two identical trains at the upstream of the Slug Catchers (SC) where most of the associated gas is separated and sent to the fuel gas system.
The liquid from the SC is heated through heat exchangers and direct fired heaters so it gets to the right temperature to release enough light ends (gas) to meet the RVP spec in the High Temperature Separators (HTS). The vapor released in the HTS is cooled down so the diluents fraction that evaporated in the HTS is condensed, separated from water and returned to the export pump suction. The remaining gas is sent to the fuel gas system and the water is sent to treatment.
The new PWTP is located at Main Station and it is designed to process 1325 Sm^3/h (200 KSBPD) of production water. Treated water has oil content below 30 ppm and total solids content below 20 ppm making it suitable for re-injection.
The PWTP consists of the following water clarifying equipment:

- Skimmer Tanks T-3900 C/D and associated Gas Boots D-3900 C/D (working in parallel), with associated pumps: Skimmer Tank Transfer Pumps, P-3901 A/B/S (P-3901 A for T-3900 C and P-3901 B for T-3900 D; P-3901 S is a common spare) and Oil Recovery Pumps, P-3951 A/B/S (P-3951 A for T-3900 C and P-3951 B for T-3900 D; P-3951 S is a common spare).
- Flotation Units X-3900 B/C with associated Flotation Transfer Pumps, P-3921 A/B/S (P-3921 A for X-3900 B and P-3921 B for X-3900 C; P-3921 S is a common spare).

- A Flotation Overflow Separation Tank T-3930 B, with associated pumps: Flotation Recycling pumps P-3926 A/S and Flotation Sludge Recovery Pumps P-3961 A/S.
- A Clean Water Tank T-3920 B to store filtered water and,
- High - pressure Water Re-injection Pumps P-3932 A/B/S.

Utilities such as Fuel gas, Firewater, Instrument Air, Plant Air, Drinking and Utility Water, Diluents required by Train C and PWTP are provided by Main Station.
The Fuel Gas used in Main Station it is provide by the incoming fluids from clusters through Trains A/B/C (Gas Oil Ratio between 50 - 250 SCF/BBL, about 12,3 MMSCFD – 61,5 MMSCFD), and the one coming from the Gas Gathering Network, for a maximum gas flow rate of 120 MMSCFD, according to gas production forecast by MSUP Study.
In order to handle this peak flow rate new Fuel Gas facilities and LP Flare System will be enhanced.
All this gas is sent to the Fuel Gas KO Drums, D-3201 A/B/C, and Fuel Gas Filters F-3201 A/B/S, and then it is distributed to the users, namely:

- Oil Heaters, H-3001 A/B/C,
- Power Generation Units, GT-3201 A/B/C/D,
- Off-spec Storage Tank, T-3002 and the Production Water Treatment Plants for blanketing purposes.
- Flare system (LP and LLP), for continuous purge.

Gas for turbogenerators is compressed in the Fuel Gas Compressor Packages, K-3201 A/B/C/S while the heaters and relief system can use it at the pressure level supplied by the KO drums.
Firewater, Instrument Air and Utility Air, Utility Water and Drinking Water connection points to Main Station are listed in Tie-in Point List. Electrical power is supplied by Turbo Generators, GT-3201 A/B/C, for MSUP a new Power Generation Unit, GT-3102D, will be installed, with its associated equipment (new generator substation G/S-02 plus an electrical substation T/S-04 dedicated to New Oil and Water Processing Trains). Additionally MS has two Emergency Diesel Generators, EDGS-1/2.
The PWTP normal operation is manual but it can be monitored from the Main Station Control Room and by sample testing at different points throughout the Plant.

7.4 Risk Analysis - HAZOP

The key to the Process Hazard Analysis (PHA) is a simple concept offers ways to answer the following four (04) questions, also a greater risk reduction, beyond the goals above, may be appropriate where an analysis cost-effectiveness of permits:

a. What can go wrong?
b. What are the causes?
c. What are the consequences?
d. What is the probability of occurrence?

The implementation of the entire set of techniques PHA comprises a quantitative review risks (Train C, New Water Treatment Production Water (NWTPW), Generation System and Common Services including here: Fuel Gas System, Low Low Pressure and Low Pressure Relief System, Diluent System, Open and Closed Drain Systems, Instrument and Plant Air System, Potable and Service Water System, Chemical Injection Systems) that may be in an installation, ranging from frequent incidents low-stakes to major incidents rare and, using a methodology uniform and consistent.

Once identified risk processes, the techniques can PHA help focus control studies risk. The PHA must address the following:

- The hazards of the process
- Identification of any previous incidents with had a likely potential for catastrophic consequences
- Engineering and administrative controls applicable to hazard, their interrelationships and consequences of failure of these controls, especially those affecting employees and environment
- Facility siting
- Human factors
- Evaluation of effect of failure of controls
- The requirement to promptly resolve PHA findings and recommendations

The largest contributors to risk can be identified and can be recommendations and take decisions, as corrective measures, on the basis objective and consistent.

The business value of the hazards analysis is the use of report to train people on how deviations from normal operation propagate to process hazards. The risk analysis defines the risk reduction strategy selected to address unacceptable risk. Plant management, engineering, operations, and maintenance must understand this information if they have responsibility for decisions affecting the hazards or risk.

Operator and maintenance procedures should include a description of the process hazards where their actions potentially increase risk. The hazards analysis documents should be discussed at Operations' safety meetings.

Residual risk should be addressed with compensating measures and action plans to reduce risk as required.

In this case, don't was identified the hazard about:

- Wrong specification for suction and discharge valves at Fired-direct Furnace, H-3001C, (Figure 8)
- Lack of control for valves at inlet pipelines
- Among others

Figure 8. Fired direct Furnace H–3001C

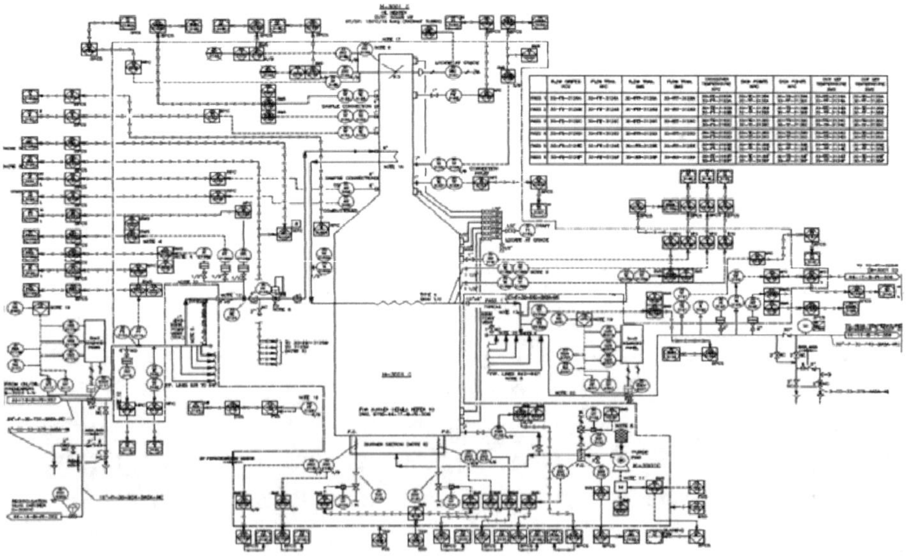

7.5 Allocation of Requirements of Safety Integrity Level (SIL) of Safety Instrumented System (SIS)

Due to the worldwide major accidents, most industrialized nations established process hazards control regulations. Germany passed the Hazardous Incident Ordinance (1980) and the European Community created the Major Accident Hazards Directive (1982). The United States passed legislation creating the Emergency Planning and Community Right-To-Know Act (EPCRA, 1986). In 1990, the Clean Air Act Amendments mandated that the Occupational Safety and Health Administration (OSHA) and the Environmental Protection Agency (EPA) develop accident prevention regulations, leading eventually to the 1992 OSHA Process Safety Management (PSM) and the 1996 EPA Risk Management Program (RMP) regulations.
None of the regulations were prescriptive. Most were intended to be flexible, simply requiring industry to examine the risk posed by their processes and to work to minimize these risks. Only a few European regulations provided quantitative risk targets. Many European regulations, as well as those of the U.S., relied on industry to develop its own tools to identify risk and to make appropriate efforts to achieve lower risk. As an incentive, industry was required to communicate the identified process risks to the public and emergency responders through public forums and documentation.
The industrial community responded to the regulations by issuing industry standards and guidelines concerning the evaluation of risk and the design of safety systems. The American Petroleum Institute (API), the American Society of Mechanical Engineers (ASME), and the National Fire Protection Agency (NFPA) created codes and practices for specific applications. In 1988, the International Society for Measurement and Control (ISA) began an eight year odyssey to develop a standard for safety instrumented system design (SIS) for the process industries, ANSI/ISA 84.01-1996; late In the mid-1980s, the International Electrotechnical Commission (IEC) began the development of an international standard for the design of all safety-related systems, covering transportation, medical, manufacturing, and process industries as IEC 61508. And then at 2003, ANSI/ISA.S.84.00.01-2004 and IEC 61511.
Consequently, it is regulations the mandate that industry identify and address risk. Industry standards are simply providing the tools for assessing the adequacy of industry's efforts. The difficult task of creating one standard for all of the process industry made the creation of a prescriptive standard impossible. After all, could one prescriptive standard cover refinery furnaces, chemical reactors, pulp and paper digesters, and utility boilers and reduce risk appropriately at a reasonable cost? Thus, the standards had to be performance based. Both standards chose to rely on the establishment of a design process called the SIS lifecycle, throughout which the performance of the instrumented systems must be maintained.
The lifecycle is intended to address the primary causes of control system incidents. Compliance with the industrial standards requires four essential elements:

a. Identification of safety functions required for safe shutdown;
b. Assignment of a safety integrity level (SIL) for each safety function;
c. Use of the safety lifecycle for the SIS design; and
d. Verification of the SIL achieved for each safety function.

These elements result in a major paradigm shift for SIS design. While it has been accepted practice for many years to mitigate potential incidents with instrumented systems, there has generally been no assessment of what type of SIS provides the appropriate risk reduction. These standards have now established requirements that industry document the rationale behind the use of a particular design, evaluate that design for its integrity, and demonstrate that the integrity can be maintained. This is perhaps the most significant impact of ANSI/ISA 84.01 and IEC 61508 on process safety, in general.

As performance based standards, the race for compliance is marked with the broadly defined finish line of good engineering practice and highly recommended practices. As industry races toward compliance, it must work hard to prevent the creation and acceptance of bad engineering practices, which threaten the economics of plant operation and erode the effectiveness of SIS designs, addressing seven bad engineering practices:

1. Believing that, if something is not specifically stated, either "shall do" or "shall not do," in the standards; you do not have to worry about it.
2. Thinking that meeting the minimum requirements means the process is safe and the SIS is compliant with the standard.
3. Ignoring the importance of good engineering practice.
4. Designing systems that meet safety requirements but not economic protection requirements.
5. Focusing only on SIL and not on preventing nuisance trips.
6. Neglecting the human factors.
7. Focusing on capital cost and not lifecycle costs.

A special attention should be paid to the possible common mode causes which may induce a failure of several (or all) barriers of the protection system. They include namely:

- The power supplies of the system.
- The logic control system(s)
- The valves activation system(s)
- The corrosion/erosion/plugging effects of the crude fluid.
- The operation of the system.

The Active Protection System is to be based on different independent barriers acting at different levels in case of an escalation of any deviation from the normal operation:

- Alarm: provides a warning to the operators in order to take a corrective action, given the deviation is not expected to become hazardous during the performance of the action.
- Production Shutdown System (PSDS): avoids the occurrence of an accidental event upon detection of a hazardous situation. The related action induces generally the shutdown of the related equipment through the implementation of one or several of the following barriers:
    - Primary barrier: aims at suppressing the cause of the deviation.
    - Secondary barrier: aims at limiting the effect of the deviation onto the equipment.

The PSDS may include several sub-levels, given there is General production Shutdown level initiating the shutdown of the whole production equipment.

Emergency Shutdown System (ESDS) triggers the safety shutdown of the installation upon detection of, either an **accidental event** or of a hazardous situation which may escalate rapidly. The ultimate ESD level causes the isolation of the installation at its limits through ESDV's and SCSSSV's (wells). It initiates the General production Shutdown and the shutdown of the related utilities. It keeps the necessary resources for emergency response. The ESDS may include several sub-levels, given the ultimate level, as defined here above, has priority on the other levels.

The Industry Practice regarding the control of the overpressure risk is to consider, by order of priority, the following possible options:

1. Design the equipment and/or the pipeline system to sustain the maximum possible pressure which may occur in the equipment (safety by design).
2. Design a protection system which, in accordance with API RP 14C, is made of:
   - A primary barrier aiming at isolating the heated equipment from the source of overpressure (suppressing the cause).
   - A secondary barrier aiming at relieving the pressure through a suitably sized venting or flaring system (limiting the effects).
3. Design an instrumented protection system made of two independent barriers acting at different levels to isolate the heated equipment from the source of overpressure before the pressure exceeds the maximum allowable pressure. This option named IPPS is considered when it is shown the options (1) and (2) induce disproportionate costs, technical or environmental problems. When a rapid escalation of the overpressure is expected, the IPPS may trigger the closure of the SCSSV's and/or ESDV's through the ESDS, in that case, the ESDS provides an additional and independent barrier.

The system should be designed fail-safe which means that the system will revert to a predetermined safe state in the event of failure of its components or of its power supplies. These include:

- Loss of the actuation energy of the valves.
- Loss of the control signal of the valves.
- Loss of the overpressure detection sub-system.
- Loss of the logic sub-system.

The overpressure protection system is to be based on the concept of two independent active barriers automatically initiated, at different levels, by the early detection of the pressure built-up into the protected equipment.

The system should survive any single failure of its components without jeopardizing its protection function. This is to be implemented through the research and elimination of the possible common mode failures, implementation of redundancies or by fail-safe action. Consideration should be given in the design of the redundancies to the using of different technologies (electrical/hydraulic valve activation as example).

The system has to be designed in order to reduce the probability of unrevealed failures of its components. This is achieved by implementation, where applicable, of monitoring and/or automatic testing of the components triggering an alarm in case of detected failure.

The system has to be designed in order to facilitate the periodic testing of the system. Consideration should be given to the possibility of performing a partial closure and re-opening of the valves while producing when this improve the availability of the protection system.

The overpressure protection system including the dedicated IPPS components should be considered as a whole system regarding its dedicated target and general objective. The reason of this approach is to facilitate the control of the overpressure hazard by considering all possible causes and their effects and namely to tackle the possible common mode failures which might involve a failure of the whole system, for example, for fired-direct furnace H-3001C it is considered an overpressure protection system that don't operate because the shutdown valves (input and output; continuing service) are specified as "Fail Last Position" and then, occurring an event will not protect the equipment, operators, assets and environment.

7.5.1 Evaluated Systems as New Train C, New Water Treatment Production Water (NWTPW), Generation System and Common Services including here: Fuel Gas System, Low Low Pressure and Low Pressure Relief System, Diluent System, Open and Closed Drain Systems, Instrument and Plant Air System, Potable and Service Water System, Chemical Injection Systems

In general, SIL3 is configured on all systems which is not sufficient safety to conduct a risk that has negative consequences and high exposure to high toxic concentrations.

Provided the single safety function (SIF) as HAZOP results, you're talking about lies within the scope of IEC 61511/62061 (i.e., a SRECS for instruments and machinery), and – as long as it functions properly – is sufficient to reduce the residual risk to an acceptable level, a reliability level of SIL3 for that single SIF is sufficient. By design that single SF will have to include sufficient redundancy/diversity/monitoring to be able to achieve SIL3.

With I mean that a if the single SF concerns e.g. the guarding of a shutdown valve to a danger zone, no other means of access to that same danger zone should exist, unless these are also guarded by SIL3 safety functions.

Remember that according to the principle of integrated safety for equipments, a SIS should only be considered "after" the possibilities of design changes and of physical protection measures have been exhausted. If the residual risk at the end is still too high, should consider additional constraints on the intended use of the equipments.

If use a simple pressure switch (diaphragm to lever-arm to electrical switch) in a safety interlock circuit where PL_a is called for, is a standard off-the-shelf pressure switch appropriate?, and If PL_c is called for (due to higher pressure), is still; possible to use a high reliability pressure switch, or would one use a pressure transmitter in a SIL1 safety function.

Can use a commercially off the shelf pressure switch provided it is a well tried component with known failure modes for a PL_a safety related control function, however to achieve PL_c, you will need to use redundant pressure switches. Using a single pressure switch may not give you the desired reliabilities due to HFT = 0 using IEC62061 or even with the use of ISO13849-1 due to a single channel circuit. However, it is important to bear in mind that the safety related control function extends from the sensing switch right up to the actuating device; and the lowest PL or SIL in the chain will influence the final result for the entire safety loop. A pressure vessel would fall under the pressure equipment directive 97/23/EC and the IEC 61511 functional safety standard for such systems would take over. Principles are similar using SIL too.

PL_c capability can be achieved in a single channel subsystem providing that the components have high $MTTF_d$. For a SIL1 according to Table 5 in IEC 62061 it is possible to achieve SIL1 with HFT = 0 providing that SFF is >60%. Single channel sensor subsystems can also achieve SIL1 according to Table 6 in IEC 61511.

The common cause and systematic failures that affect even a SIL Logic Solver will affect all protection layers that the logic solver is part of, and these will as a consequence of this not be independent. In fact this is subject to discussion, where some are of the opinion that a SIL3 logic solver is able to cover both the 2 IPLs required by ISO 10418.

These SIFs would not qualify to be IPLs. However, I would think that the SIFs should reinforce the protection towards the same risk, especially if the Logic Solver has very low failure rate compared to the field devices.

The independent rule would be hurt. According Layer of Protection Analysis, CCPS 2001: "In order to be considered an IPL, a device, system, or action must be independent of the initiating event and the components of any other IPL already claimed for the same scenario…" However, one of them could be considered as a safeguard, none SIL claimed.

Depends on having independent paths through the SIL3 Logic Solver. It is possible to have two separate SIF's with different initiators and different final elements to cater for the same risk but same logic solver (SIL3 rated) are these IPL's.

Adding SIFs for the same risk don't provide you with more IPL because:

- They aren't independent, because even they use different instruments; they share the same logic solver.
- They aren't different IPL, because both of them are SIF. An extra PL could be a mechanical relief valve, and plus, it will be independent from the logic solver, so you will have another IPL. That way if the inner previous IPL doesn't work (you can consider your SIF as this inner IPL) you will have another extra IPL.
- The common cause and systematic failures of the SIL3 Logic Solver will affect both protection layers. Therefore the LS IPL load is limited to 3, one could load a 2 SIL1 loops or 1 SIL1 + 1 SIl2 on this system, but one could NOT load 2 SIL2 loops on this system. This according IEC 61511-part 1, clause 9.5.1, part 2 Clauses 9.5.3.
- As you know that SIS consists of SIF's and you can use single or redundant logic solver for those SIF's (in SIS). So my answer for using same logic solver for many SIF is yes you can, single/redundant LS will be depend to your verification that you will count many things (diagnostic covered, beta factor for common cause, etc) to make sure that your SIF equipment reliability will meet level target (SIL target / PFD).

A Common Cause Failure (CCF) occurs when a single fault results in the corresponding failure of multiple components, such as a miscalibration error on a bank of redundant transmitters. The frequency of common cause faults is difficult to estimate. The modeling techniques and available failure rate data make the predictive calculations of these failures cumbersome and the results obtained questionable. Therefore, a more meaningful way for most SIS designers is to eliminate the potential source of CCF in the SIS design, installation, operation and maintenance.

The SIS should be designed to meet the required safety integrity level as defined in the safety allocation requirement specification. Moreover, the SIS design should be performed in a way that minimizes the potential for common mode or common cause failures. A CCF occurs when a single fault results in the corresponding failure of multiple components. Thus, CCF's can result in the SIS falling to function when there is a process demand. Consequently, CCF's must be identified during the design process and the potential impact on the SIS functionality must be understood.

Unfortunately, there is a great deal of disagreement among the experts on how to define CCF and what specific events comprise a CCF. The following are often cited as examples of common cause faults:

- Miscalibration or no calibration of sensors
- Pluggage of common process taps for redundant sensors
- Incorrect maintenance or no maintenance
- Improper bypassing
- Environmental stress on the field device
- Process fluid or contaminant plugs valve

But the examination of these faults, in light of any SIS design, will indicate that any of these six examples can disable single I/O systems, as well as redundant I/O systems. However, many of the proposed methodologies for assessing CCF ignore this fact and only penalize redundant sensors and final elements.

The ultimate failure of all is that the safety requirement specification (SRS) is incorrect at the beginning of the design process and the transmitter cannot detect the potential incident. This is the most disastrous common cause failure that directly leads to the hazardous incident that the designer is seeking to prevent. A bad SRS is the ultimate "trump card" for the entire SIS and is one failure that most of the proposed methodologies ignore.

The choice of the evaluation technique is typically dependent on experience of the User with the particular SIS design. This would include documented historical performance of instruments, installation details, and design engineering teams. Experience with the specific application environment is required, because a device or installation detail that works well in one type of application may not work well in another. For example, standard taps into a vessel for mounting a transmitter may work extremely well in clean service, but may plug very quickly in a service where solids can deposit. Three qualitative techniques are often used to assess SIS designs:

- Industrial Standards
- Engineering Guidelines and Standards
- Qualitative Assessment as Checklist

8. Conclusions and Recommendations

These new rules require a change of focus for everyone associated with these systems. The change of focus will result in a significant learning curve for many engineers. In order to make this transition as smooth as possible, rigorous qualitative techniques should be employed to provide a method for assessing these systems for adequacy. Clear communication of the requirements for effective SIS design is essential. All procedures, internal guidelines, and checklists must be evaluated by all personnel involved to ensure a compliant and cost effective SIS.

One suggestion that is obvious but difficult due to human nature is – help management (refinery/corporate) to truly see the OSHA inspection as a positive learning opportunity.

An efficient, accurate OSHA inspection can provide direction to upgrading a refineries PSM program into one that supports sustainability. Long term maintenance of a rigorous refinery PSM system will prove its business case in terms of:

- Corporate Responsibility
- Business Flexibility
- Risk Reduction
- Sustained Value

An effective PSM system upgrade resulting from the inspection can actually increase a refinery's reliability and ability to sustain long term success.

The refining industry will have a directed opportunity to improve overall performance in process safety due to the OSHA refinery process safety management national emphasis program. Leveraging this opportunity requires a culture that views personnel and process safety as second nature – a conditioned first response to address for any issue related to refinery operations or maintenance.

High reliability organizations (such as refineries, petrochemical complexes, chemical plants and power plants) whose upper management values long term growth and more predictable physical plant reliability will embrace the chance to use the effort involved in a PSM program implementation based upgrade to bolster these benefits.

We now have benefits and costs. How are these converted into a proper overall project financial analysis? How can we demonstrate that the proposed investment will increase the long term corporate ROIC more than other potential investments?

The cultural level of operational discipline at the corporation, team, and individual level can better support a high level of operational excellence – doing those critical measurable things that support success – when a program is detailed and monitored. Business excellence can be attained when a company values a holistic sense of what comprises quality.

The Figure 9 represents this view. OSHA inspectors and management have gained a new level of process safety management expertise based upon their roles in the Industry. This can only enhance OSHA's ability to help industry protect the workforce from preventable risk. If production facility management adopts this view, the industry's level of catastrophic risk prevention will increase.

Figure 9. The Holistic View of Quality

Although the specific investment evaluation protocols vary from company to company and you need to follow your company's standards, they all have at their center the calculation of the net present value of the incremental after tax benefits generated by the plant due to the equipment or technology and comparing it with the net present value of the required investment. In financial terms, all assets, be they stocks, bonds or plants, are valued the same, i.e. as the net present value of the future sequence of after tax cash flows. If the value is greater than the investment the project is considered for funding. Net present value implies discounting a sequence of net cash flows (receipts less expenditures) back to the current time. But there are many subtleties.

The Operation and Maintenance staff of today must justify how the SIS is operated, when it is bypassed, when it is tested, and when maintenance is performed.

9. Acknowledgement

The author would like to thank our Meeting assistant colleagues for many discussions, which assisted in the evolution of the ideas presented herein to:

AIChE's 2011 Annual Meeting, United States of America, October 16-21, 2011, Minneapolis, MN, and

inelectra Operations VicePresidency (Eng. Milagros Sulbarán) and Engineering Management (Eng. Victoria Granados), for their cooperation and support.

11. Author: Glenda J. Gutiérrez Capriles

She was born in Valencia, Carabobo, Venezuela on January, 1974. She obtained her B.Sc. from Los Andes University (ULA) in Merida - Venezuela, in 1996. Then in 1999-2000 obtained her title as Magisterium Scientarium Specialist in Environmental, Health and Safety, ULA.

Also she holds certification in PMP and TUV FS.

Between 2000 – 2008 she served from Process Specialist, Instrumentation Engineer, Risk Engineer, Maintenance Centered in Reliability Supervisor, Operational Engineering Superintendent for PDVSA E&P, PDVSA GAS and PEQUIVEN.

Gutiérrez Capriles is currently worked as Project Risk Engineer for inelectra Venezuela from beginning of 2009.

Gutiérrez Capriles has over fourteen (14) years of experience as Engineer, Consultant, Operator, highlighting those regarding Risk Studies as Inherently Safer Designs, HAZID, HAZOP, SIL/SIS, FTA, ETA, FMEA, Reliability Analysis as RCM, RBI, OEE, RAM, RAMOS or MORAS, Project Risk Analysis as LCC, DSD, PEP, among others.

She is currently worked as Project Engineer for inelectra Venezuela from the beginning of 2009. She has over fifteen (15) years of experience as engineer, consulter, operator, highlighting those regarding risk studies, reliability analysis, among others.

Her experiences in Process Safety Knowledge are cited on many national and International Journal and Seminars. She has published and presented papers at several events organized by relevant institutions at national and international levels.

- 2010 Chemical Engineers Argentine Congress, Mar del Plata, Argentina.
- 3rd CCPS Latin America Conference and Expo, Buenos Aires, Argentina.
- 2011 Annual Meeting AIChE, Minnesota, USA.
- AIChE Journal, March 2001 Performance Assessment of Constrained Model Predictive Control Systems.
- Gutiérrez-Capriles G. "I Jornadas de Protección Integral". Universidad de Los Andes. Facultad de Ingeniería. Mayo 1996.
- Gutiérrez-Capriles G. Ergonomía en los diferentes Ambientes de Trabajo (Tesis de Magister). Mención Especial de Honor por la Sociedad Venezolana de Medicina Ocupacional, Escuela de Salud Ocupacional de la Facultad de Medicina y Escuela de Protección Integral de la Universidad de Los Andes. Universidad de Los Andes. Mérida, Venezuela (1999)
- Gutiérrez-Capriles G. Anteproyecto del Diseño de una Planta para la Obtención de Lecitina Blanqueada de Soya (Diseño de Plantas). Memorias del II Congreso Nacional de Estudiantes de Ingeniería Química. Universidad de Los Andes. Mérida, Venezuela (1999)
- Gutiérrez-Capriles G. Peña A. BATCH: Programa de Computación simulando una Torre de Destilación por Lotes Batch. Memorias de las IV Jornadas Científico-Técnicas de la Facultad de Ingeniería. Universidad de Los Andes. Mérida, Venezuela (1998)
- Gutiérrez-Capriles G. "III Congreso Internacional de Educación Superior en materia de Drogas". Universidad de Los Andes. Abril 1996.
- Gutiérrez-Capriles G. "I Congreso Internacional de Educación Integral en materia de Seguridad sobre Riesgos Sísmicos". Vicerrectorado Académico y Brigada de Rescate de la Universidad de Los Andes. Abril 1996.
- Gutiérrez-Capriles G. Optimización de Mezclas de Policloruro de Vinilo utilizadas para la Elaboración de Envases Plásticos (Tesis de Grado). Mención Especial de Honor por la Sociedad Venezolana de Ingenieros de Polímeros, Departamento de Química de la Facultad de Ciencias y Departamento de Ingeniería Química de la Universidad de Los Andes. Universidad de Los Andes. Mérida, Venezuela (1996)
- Gutiérrez-Capriles G, Belandria J. Graficador Termodinámico de Compuestos Químicos Puros en Fase Fluida por 4 diferentes Métodos Termodinámicos. II Congreso Interamericano de Computación Aplicada a la Industria de Procesos. Santiago de Chile, Chile (1994)

(355a) Liquid Crystal Based Optical Sensors for Detection of Aliphatic Amines In Air

Xiaokang Ding and Kun-Lin Yang, Department of Chemical and Biomolecular Engineering, National University of Singapore, Singapore, Singapore

Liquid Crystal Based Optical Sensors for Detection of Aliphatic Amines in Air

Prompt detection of aliphatic amines is an important task in environmental monitoring of air pollutants and quality control in food industry. In this presentation, a new type of liquid crystal (LC) based optical sensor with fast bright-to-dark response to butylamine in vapor phase will be discussed. We first report orientational behaviors of a thin film of lauric aldehyde (LA) doped liquid crystal, 4-cyano-4′-pentylbiphenyl (5CB), in three different configurations (air/LC/air, glass/LC/glass, and air/LC/glass) and study the effect of LA on the orientational behaviors 5CB at different interfaces. Our result shows that at the LC/air interface, the LC orientation is homeotropic when the LA concentration is below 0.5 wt%. In contrast, at the LC/glass interface, the LC orientation is homeotropic when the LA concentration is above 0.5 wt%. Moreover, because LA is reactive to butylamine, when we exposed 0.1 wt% LA-doped LC to 10 ppmv of butylamine vapor, the LC undergoes a rapid bright-to-dark optical response within 2 min. This response is caused by the orientational transition triggered by the reaction between LA and butylamine. In contrast, undoped LC did not show such response. This study shows the active role of LA dopant in determining the orientations of LC at two different interfaces, and its potential application in developing LC based gas sensors.

See more of this Session: Catalytic, Environmental, and Industrial Gas Sensors
See more of this Group/Topical: Topical 9: Sensors

(76f) Surface Plasmon Resonance (SPR) Biosensors for Specific Detection of Neonicotinoids

Xiaokang Ding and Kun-Lin Yang, Department of Chemical and Biomolecular Engineering, National University of Singapore, Singapore, Singapore

Surface Plasmon Resonance (SPR) Biosensors for Specific Detection of Neonicotinoids

Neonicotinoids such as thiacloprid and imidacloprid are a class of pesticides which are able to block nicotinic acetylcholine receptors in the central nervous system of insects. In this presentation, we first introduced two highly specific oligopeptide sequences RKRIRRMMPRPS ($\mathbf{P_1}$) and RNRHTHLRTRPR ($\mathbf{P_2}$) as biomolecular receptors for binding neonicotinoids such as thiacloprid and imidacloprid. These two oligopeptides were identified after 3 rounds of panning processes by using phage display library. The former oligopeptide $\mathbf{P_1}$ shows high affinity for thiacloprid whereas the latter oligopeptide $\mathbf{P_2}$ shows high affinity for imidacloprid. By immobilizing these two oligopeptides on SPR sensor chips, we successfully prepared a biosensor system for on-line monitoring of either thiacloprid or imidacloprid in buffer solutions. Our SPR experiments confirm the strong binding thiacloprid and imidacloprid to surface immobilized oligopeptides. The limit of detection for thiacloprid and imidacloprid is approximately 2 μ M. Surprisingly, despite the similarity of thiacloprid and imidacloprid, cross-binding is minimal. Experimental results for analyzing neonicotinoids on fruits and vegetable samples will also be discussed in this presentation.

See more of this Session: Biosensor Devices II
See more of this Group/Topical: Topical 9: Sensors

(13f) Reforming of Residual Tars and Oils From Biomass Gasification

Lyman Frost[1], Elango Elangovan[1] and J. Hartvigsen[2], (1)Ceramatec, Inc, Salt Lake City, UT, (2) Electrochemical, Ceramatec, Inc, Salt Lake City, UT

One of the barriers to wider use of many kinds of biomass is the fact that they generate refractory tars and oils when gasified. These compounds cause difficulty when they are passed to downstream components in the system by causing plugging and drop in performance. The compounds can be condensed and removed but this involves extensive heat exchange, increased cost, and removal to a hazardous waste landfill.

Working as a subcontractor to Emery Energy, a Salt Lake City company with expertise in biomass gasification, Ceramatec will be applying its non-thermal plasma reformer technology to eliminate any refractory tars and oils that exit the Emery gasifier. Preliminary laboratory work indicates that it is possible for the plasma reformer to operate on the gas exiting the gasifier and convert these tars and oils to additional hydrogen and carbon monoxide. This work is being done at Western Research Institute in Laramie, Wyoming under a US Department of Energy award to Emery Energy.

This paper will present an overall process structure of the gasification process and the non-thermal plasma reformer in particular. Operational characteristics and laboratory results related to the reforming of the refractory tars and oils will be presented. The larger pilot sized facility at Laramie will also be described. Data is expected to be available from the larger pilot sized facility in Laramie by the time for the conference. The resultant gas stream is of a quality that could be used by a high temperature fuel cell.

See more of this Session: Catalytic Conversion of Renewable Resources to Synthesis Gases and Pyrolysis Oils
See more of this Group/Topical: Sustainable Engineering Forum

(355b) ZnO-Based Thin Films for Applications In Chemical Sensing

J.B. Miller, T. Ashok, S. Lee and E. Broitman, Chemical Engineering, Carnegie Mellon University, Pittsburgh, PA

Zinc oxide (ZnO) is a wide band-gapsemiconductor with important applications in electronics, photonics, chemicalsensing and catalysis. Sol-gelchemistry provides a convenient route for low-cost fabrication of ZnO functionalcoatings with controlled microstructure. Critical textural, chemical andelectronic properties can be further manipulated by doping ZnO with minorcomponents. In this work, we applied sol-gel wet-chemical techniques topreparation of ZnO, Al-ZnO (Al:Zn = 1:10) and Cu-ZnO(Cu:Zn = 1:10) thin film functional layers for chemiresistive sensors.

Cu and Al dopants influence the films'surface morphology and their thermally induced chemical and structuralevolution. As prepared (room temperature) films exhibit the structure ofLayered Basic Zinc Acetate (LBZA), a lamellar ZnO precursor. When annealed at temperatures through700 °C, the films display similar chemical evolutionpatterns—characterization by IR spectroscopy reveals that temperaturesabove 500 °C are needed to completely desorb solvents and decomposeprecursors. X-ray diffraction results show that Cu facilitates c-axisorientation of the annealed film, while Al slows its crystallization. Detailsof surface morphology, imaged by electron microscopy, also depend on the choiceof dopant.

UV-vis spectra show that the Al dopantslightly widens the band-gap of calcined film, while Cu increases absorptionbroadly throughout the visible range. Chemiresistive sensors, fabricated bycoating thin film functional layers onto interdigitated electrode transducers,were evaluated for their sensitivity to oxygen at operating temperaturesthrough 600 °C. Asensor coated with undoped ZnO displays good sensitivity for O_2 atintermediate temperatures, ~400 °C, likely reflecting an optimalbalance between surface O_2 coverage and carrier availability. Cu-ZnO, on the other hand, has a higherbase resistance (in nitrogen) and is less sensitive to O_2.

See more of this Session: Catalytic, Environmental, and Industrial Gas Sensors
See more of this Group/Topical: Topical 9: Sensors

(471b) Mesoporous TiO2 Beads From a Facile One-Step Template-Less, Surfactant-Free Hydrothermal Process and Derived TiO2 Bead/ Xerogel Composites for High Efficiency Dye-Sensitized Solar Cells

Wei-Yung Cheng, Juti Rani Deka, Yi-Chun Chiang, Antoine Rogeau and **Shih-Yuan Lu**, Dept. of Chemical Engineering, National Tsing-Hua University, Hsin-Chu, Taiwan

A facile one-step template-less, surfactant-free hydrothermal process is developed to produce sub-micron sized mesoporous TiO_2 beads of high specific surface areas. No templates or structure directing surfactants are involved in the process. In addition, the beads are produced in one-step, unlike the two-step process (sol-gel followed by solvothermal) commonly adopted in the literature. The as-prepared TiO_2 beads are crystalline of the anatase phase, with a high specific surface area of 150 m^2/g and an average pore size of 7.8 nm, and sub-micron in size, 500-800 nm. TiO_2 xerogels are incorporated into the TiO_2 bead layer to form a TiO_2 bead/xerogel composite to serve as the photoanode of dye-sensitized solar cells. A high power conversion efficiency of 8.41% is achieved from the TiO_2 bead/xerogel composite photoanode, representing a 33% improvement over the efficiency of 6.32 % achieved with a P-25 TiO_2 photoanode under the same characterization conditions. The success of the present composite TiO_2 nanostructure can be attributed to the excellent structure connectivity within the bead and xerogel domains for easy electron transport, the high specific surface areas of both the beads and xerogels for providing abundant dye adsorption for generation of photon-induced electrons, and the submicron size of the beads for much improved light harvesting efficiency resulting from the intense light scattering.

See more of this Session: Nanostructured Materials for Dye-Sensitized Solar Cells
See more of this Group/Topical: International Congress on Energy 2011

(13a) Mechanistic Modeling of Fast Pyrolysis of Cellulose to Predict Bio-Oil Composition

Vinu Ravikrishnan and Linda J. Broadbelt, Chemical and Biological Engineering, Northwestern University, Evanston, IL

Fast pyrolysis, a potential strategy for the production of transportation fuels from biomass, involves a complex network of competing reactions, which result in the formation of bio-oil, non-condensable gaseous species, and solid char. Bio-oil is a mixture of anhydro sugars, furan derivatives, and oxygenated aromatic and low molecular weight (LMW) compounds. Currently, the successful modeling of fast pyrolysis reactors for biomass conversion is hampered by lumped kinetic models, which fail to predict the bio-oil composition. Hence, a fundamental understanding of the chemistry and kinetics of biomass pyrolysis is important to evaluate the effects of process parameters like temperature, residence time and pressure on the composition of bio-oil.

We have recently developed a comprehensive mechanistic model to characterize the primary products of fast pyrolysis of pure cellulose. The model incorporates the following condensed phase reactions with ionic and electrocyclic fragmentation mechanisms for the formation of various LMW products in bio-oil: glucosidic bond scission, retro-aldol, retro-Diels-Alder, 1,2-dehydration, 1,3-dehydration, hydrolysis, cyclic Grob fragmentation and enol-keto transformation reactions. The kinetic rate coefficients for most of the above reaction steps were obtained from the literature, based on either experimentally determined Arrhenius parameters for cellulose and model compounds, or theoretically computed values of activation energies using quantum chemical calculations. A computational framework based on continuous distribution kinetics was constructed to solve the kinetic model. The model tracks 96 distinct polymer species and 25 LMW products, and incorporates over 350 reactions. The model predictions compare well with the experimental data (Patwardhan *et al.*, *J. Anal. Appl. Pyrolysis* **2009**, 86, 323-330) for the formation of important products like levoglucosan (60 wt.%), glycolaldehyde (6 wt.%), 5-hydroxymethyl furfural (3 wt.%), furfural (1 wt.%) and formic acid (6 wt. %). Levoglucosan was found to be formed by the initiation of ionic chain ends by mid-chain glucosidic bond scission, followed by end-chain ionic unzipping. Some of our efforts in extending the kinetic model of pyrolysis of pure cellulose to cellulose with mineral matter and hemicellulose will also be discussed in this presentation.

See more of this Session: Catalytic Conversion of Renewable Resources to Synthesis Gases and Pyrolysis Oils
See more of this Group/Topical: Sustainable Engineering Forum

(471g) Highly Catalytic Pt Nano-Particles As An Alternative to Pt Films At the Counter Electrodes of Dye Sensitized Solar Cells

Somik Mukherjee[1], Balavinayagam Ramalingam[1], Lauren Griggs[1], Steven Hamm[1], Shubhra Gangopadhyay[1] and Shramik Sengupta[2], (1)Electrical and Computer Engineering, University of Missouri, Columbia, MO, (2)Biological Engineering, University of Missouri, Columbia, MO

Since their introduction in 1991, Dye Sensitized Solar Cells (DSSCs) have emerged as a low-cost alternative to traditional (thin film) solar cells. An important goal of current research in the area is to improve the Price/Performance Ratio. The DSSCs with the best performance use Platinum as the counter-electrode catalyst for Iodide/Tri-iodide reduction. Typically, the platinum is deposited as a thin film (~50nm for front-side illuminated DSSCs, and ~2nm for backside illuminated DSSCs).

In this work, we report an alternative that uses significantly less platinum but is able to deliver catalytic performances (efficiencies and fill factors) comparable to a 50nm thin film, and has transmittances that are better than those of 2nm Pt films.

This is achieved by depositing Platinum Nanoparticle on Fluorine-doped Tin Oxide (FTO) films in a controlled manner using RF Magnetron Sputtering at room temperature. The deposited platinum aggregates in the form of "nano-islands" on the FTO surface, with the size of the islands being primarily determined by the duration of the sputtering.

The optimum performance (photo-conversion efficiency of 4.9%) is obtained with a sputtering time of 45s. For this sputtering time, our nano-islands are approximately 2-3 nm in diameter and have a surface density of approximately 5.9 E12 per square cm. The amount of Platinum used is thus ~ 400 times less than that in a 50nm Pt film, where similar photo-conversion efficiencies are obtained.

An additional advantage of using Pt nanoparticles (as opposed to films) is that the transmittance of the layer to radiation in the UV-visible range is also greatly enhanced. Whereas Pt films typically suffer from a significant loss (absorption/reflection) of radiation (5-10% for 2nm thin films, and as much as 90% for 50 nm films), our films consisting of Pt Nano-particles on FTO show minimal changes in transmittane. This implies that our Pt-nanoparticle based counter electrodes can potentially be used in DSSCs that utilize back-side illumination as well.

See more of this Session: Nanostructured Materials for Dye-Sensitized Solar Cells
See more of this Group/Topical: International Congress on Energy 2011

(355c) Single Walled Carbon Nanotubes Based Ultra-Sensitive Resistor-Type Gas Sensors

Ying Wang[1], Yu Ding[1], Joshua Akhigbe[2], Christian Brckner[2] and Yu Lei[1], (1)Chemical, Materials & Biomolecular Engineering, University of Connecticut, Storrs, CT, (2)Department of Chemistry, University of Connecticut, Storrs, CT

Following facile synthetic procedures (electrospinning and membrane filtration), two single walled carbon nanotubes (SWNTs) based sensory materials have been prepared and applied for resistor-type gas detection. Both SWNTs based sensors have been characterized by scanning electron microscopy (SEM), transmission electron microscopy (TEM), Fourier transform infrared spectroscopy (FT-IR) and UV-Vis. Specifically, incorporation of SWNTs into PPy nanocomposites by electrospinning generates a novel SWNTs composite nanofibers and a synergistic effect between SWNTs and PPy, resulting in an improved NH_3 detection limit as low as 36 ppb. Another SWNTs based sensor has been developed by the noncovalently immobilization of *meso*-tritolylcorrole (TTC) on SWNT networks through membrane filtration. This SWNT hybrid material possesses a donor-acceptor heterojunction through the π-π stacking, which facilitates exciton dissociation and thus greatly enhances the sensitivity towards NO_2 detection. Our experimental results show great potential of novel SWNTs based sensory materials in gas detection.

See more of this Session: Catalytic, Environmental, and Industrial Gas Sensors
See more of this Group/Topical: Topical 9: Sensors

(11h) Ultra-Sensitive Glucose Detection Using Copper Nanowires

Liang Su[1], Yuchan Zhang[2], Wenzhao Jia[3], Changjun Hou[2], Danqun Huo[2] and Yu Lei[3], (1)Chemical Engineering, University of Connecticut, Storrs, CT, (2)Bioengineering College, Chongqing University, Chongqing, China, (3)Department of Chemical, Materials & Biomolecular Engineering, University of Connecticut, Storrs, CT

The detection of glucose, which is important in the diagnosis of diabetes and quality control of fermentation and food, accounts for about 85% of the entire biosensor market. In this work, both enzymatic and non-enzymatic glucose biosensors were constructed on the basis of copper nanowires synthesized by wet chemistry. The as-prepared copper nanowires were then characterized by scanning electron microscopy, transmission electron microscopy, and X-ray diffraction. In the mode of enzymatic glucose sensing, the reduction of hydrogen peroxide was first investigated. The limit of detection for hydrogen peroxide was 200 nM at -0.2 V (vs. Ag/AgCl 3 M KCl) which provides a prospective route for the detection of glucose with the incorporation of glucose oxidase. In the mode of non-enzymatic glucose sensing, an impressive detection limit as low as 50 nM with a sensitivity of 5970 $\mu A \cdot cm^{-2} \cdot mM^{-1}$ was obtained.

See more of this Session: Biosensor Devices I
See more of this Group/Topical: Topical 9: Sensors

(231a) A Novel SHF Process for Conversion of AFEX-CS to Ethanol Featuring High Ethanol Productivity, Enzyme Recycling, and Yeast Cells Reuse

Ming Jie Jin, Christa Gunawan, Nirmal Uppugundla, Shishir Chundawat, Bruce Dale and **Venkatesh Balan**, Chemical Engineering and Material Science, Michigan State University, East Lansing, MI

There are two major problems in current cellulosic ethanol production process: low ethanol productivity and high enzyme loading. Based on the current Separate Hydrolysis and Fermentation (SHF) process, it takes at least 96h to perform enzymatic hydrolysis to convert AFEX-CS to fermentable sugars at 6% glucan loading. It will then take another 96-168h to ferment those sugars to ethanol. By simply integrating enzymatic hydrolysis and fermentation, which is SSCF (simultaneously saccharification and co-fermentation), it still requires at least 168h to reach comparable results as SHF. Here we will present a novel SHF process. In this process, the easily hydrolyzed part of biomass is hydrolyzed first to fermentable sugars in 24h. The hydrolysate was then separated from residue solids and fermented to ethanol in 24h using high yeast cell inoculation. The residual solids (difficultly hydrolyzed part of biomass (DHPB)), which bound enzymes, are then transferred to the next cycle of hydrolysis. This way, enzymes are largely reused and enzyme loadings in the next cycles can thus be reduced. The yeast cells after 24h fermentation are also recovered and reused for the next cycle of hydrolysate fermentation. The time for each cycle hydrolysis/fermentation is 24h/24h. After five cycles, the difficult hydrolyzed part is collected and enzymatic hydrolysis is performed without adding additional enzymes. Since enzymes are accumulated to a relatively high concentration and part of the decomposition products are removed with liquid during the five cycles, the hydrolysis of DHPB would be easier to achieve high sugar conversion. Profiles of enzyme recycled in each step were characterized.

See more of this Session: Integrated Processes for Biochemical Conversion of Renewable Feedstocks to Fuels and Chemicals I
See more of this Group/Topical: Sustainable Engineering Forum

(491a) Glycome Profiling Using Glycan-Directed Monoclonal Antibodies: Applications to Plant Cell Wall/Biomass Characterization

Sivakumar Pattathil[1], Michael G. Hahn[2], Heather McCormick[1], Jeffrey Miller[2], Virginia Brown[2], Ajaya Biswal[2], Trina Saffold[2], Malcolm O'Neill[2], Debra Mohnen[2], William S. York[2], Jaclyn D. DeMartini[3], Charles Wyman[4], Michael W.W. Adams[5], Irina Kataeva[6], Richard A. Dixon[7], Huanzhong Wang[7] and Fang Chen[7], (1)Complex Carbohydrate Research Center, University of Georgia, Athens, GA, (2)University of Georgia, Athens, GA, (3)Chemical and Environmental Engineering, University of California Riverside, Riverside, CA, (4)University of California Riverside, Riverside, CA, (5) Biochemistry and Molecular Biology, University of Georgia, Athens, GA, (6)Department of Biochemistry, University of Georgia, Athens, GA, (7)Samuel Roberts Noble Foundation, Ardmore, OK

Plant biomass is the major source for lingo-cellulosic bioenergy. Plant cell walls form the major part of the plant biomass. Research on plant cell wall/biomass is challenging owing to the complexity of the cell walls. Thus, one of the focus areas of bio-energy research has been the development of advanced tools that enable easy, rapid, high-throughput and reliable characterization of plant cell walls/biomass. Cell wall-directed monoclonal antibodies have now emerged as powerful probes to assist in the detailed structural analysis of plant cell walls. The worldwide collection of plant cell wall glycan-directed monoclonal antibodies is now sufficiently large and comprehensive to be used to monitor the majority of cell wall glycans. Here we report an ELISA-based method, Glycome Profiling, that employs a toolkit of 155 monoclonal antibodies (recognizing most major classes of plant polysaccharides) for characterizing diverse plant cell walls. The method involves sequential extraction of cell wall materials with increasingly harsh reagents to obtain fractions enriched in diverse wall polysaccharides, such as pectins, arabinogalactans, xylans, xyloglucans, and mannans. Each of these fractions is subjected to ELISA against the toolkit of antibodies and the binding responses are depicted as heat maps. The studies presented here outline multiple applications of Glycome Profiling to plant cell wall/biomass characterization and utilization. [This work is supported by a grant from US Department of Energy BioEnergy Science Center and grant DE-PS02-06ER64304. The BioEnergy Science Center is a U.S. Department of Energy Bioenergy Research Center supported by the Office of Biological and Environmental Research in the DOE Office of Science. Generation of the monoclonal antibodies used in this work was supported by the NSF Plant Genome Program (DBI-0421683)]

See more of this Session: Advances In Biofuels: DOE Bioenergy Research Centers I
See more of this Group/Topical: Sustainable Engineering Forum

(258d) Effects of Mixing Quality On the Kinetics of Biomass Liquefaction

David M. Lavenson[1], Emilio J. Tozzi[1], Tina Jeoh[2], Michael J. McCarthy[3] and Robert L. Powell[1], (1) Chemical Engineering and Materials Science, University of Calfiornia Davis, Davis, CA, (2)Biological and Agricultural Engineering, University of Califonrnia, Davis, Davis, CA, (3)Biological and Agricultural Engineering, University of Calfiornia Davis, Davis, CA

Abstract: Biochemical conversion is a promising route to produce fuels and chemicals from cellulosic biomass. A key step in the process is biomass liquefaction, where enzymes are used to break down pretreated cellulosic particles. During this process, drastic changes in the mechanical properties of the cellulosic slurry cause a transition from a solid-like material to a pumpable liquid. A fast liquefaction is desirable to achieve reductions in equipment size and cost. To assess the effect of mixing quality on the speed of liquefaction, cellulases were added to static fiber beds in vials. Different initial mixing conditions were used. One of the conditions was a well-mixed slurry where the enzyme is uniformly distributed in the slurry. The opposite extreme condition was addition of a thin layer of enzymes on one end of the vial, representing a completely unmixed enzyme. Samples were incubated for 96h at 50C. The state of liquefaction of the fiber beds was assessed using a mechanical penetrometer test. Large differences in the force-displacement curves were observed between untreated control slurries and enzymatically treated slurries. More importantly among the treated ones, the liquefaction times were much shorter for well-mixed slurries than for poorly mixed ones. Liquefaction data was also compared with reducing sugars measured throughout the incubation period to correlate with product yields.

See more of this Session: Reaction Kinetics and Transport Fundamentals for Biomass Conversion
See more of this Group/Topical: Sustainable Engineering Forum

(339c) Effect of Storage Method and Duration On the Bioprocessing of Lignocellulosic Biomass

Arun Athmanathan, ABE/Lorre, Purdue University, West Lafayette, IN and Nathan S. Mosier, Agricultural and Biological Engineering & Laboratory of Renewable Resources Engineering (LORRE), Purdue University, West Lafayette, IN

Lignocellulosic biorefineries require an efficient system of biomass storage between harvesting and processing. This study reports the effect of storage method and duration on the composition and processing performance of sweet sorghum bagasse. Sorghum bagasse was harvested at 85% moisture content, crushed to extract the juice, and then stored. Four storage conditions were used: frozen after crushing (control), ensiled after crushing (67% moisture content), or dried in the field and baled at high (27% by weight) or low (11% by weight) moisture contents. After 24 weeks of storage indoors at 20°C, 5.5% dry matter was lost from low moisture content bales, 29.4% from high moisture bales, and 7% from silage. Compositional analysis and mass balances determined that mass loss was primarily carbohydrates. Pretreatment and SSF at high solids loading (15% w/w) were then carried out for the materials. Distinct differences in performance were noted, with lower enzymatic digestibility of the glucan observed for dry storage compared to wet storage.

To examine non-compositional impacts of storage method, bagasse samples were enzymatically hydrolyzed at 1% w/w solids loading without drying, air-dried (20°C), and oven-dried (45 °C). Release of glucose stopped after <5 hours for dried bagasse with final yields being 70% (air-dried) and 57% (oven-dried) of total glucan. By comparison, non-dried samples continued to hydrolyze beyond 5 hours, resulting 87% glucan yield after 24 hours. These results indicate that hornification during drying may also negatively affect enzyme accessibility to the cellulose and lower bioprocessing performance.

See more of this Session: Sustainable Biomass Feedstock Production and Supply for the Emerging Biorefinery Industry
See more of this Group/Topical: Sustainable Engineering Forum

(235a) Integrated Plasmonic Biosensors with Microfluidics

Peng Jiang, Chemical Engineering, University of Florida, Gainesville, FL

Surface-enhanced Raman scattering (SERS) and surface plasmon resonance (SPR) detection are two key surface plasmon techniques that will ultimately enable single-molecule-level chemical and biological sensors. To resolve the reproducibility issue of stochastically aggregated colloidal nanoparticles which are commonly used in SERS and SPR detection, various periodic plasmonic nanostructures with well-defined SP properties have been broadly exploited. Unfortunately, the development and implementation of these nanostructured SP substrates have been greatly impeded by expensive and painstaking top-down nanofabrication (e.g., electron-beam lithography and focused ion-beam), which limit the available sample size to less than 1 mm^2. Here we report a robust bottom-up nanofabrication platform that provides a much simpler, faster, and inexpensive alternative to nanolithography in creating a large variety of highly ordered plasmonic nanostructures. This platform combines the simplicity and cost benefits of bottom-up colloidal self-assembly with the scalability and compatibility of top-down microfabrication. We have demonstrated the capability of this novel methodology in creating highly sensitive and reproducible SERS and SPR biosensors over wafer-sized areas. Most importantly, this novel nanofabrication platform is compatible with standard microfabrication, enabling on-wafer integration of SP biosensors with conventional microfluidic devices.

See more of this Session: Micro and Nanofabricated Sensors I
See more of this Group/Topical: Topical 9: Sensors

(491d) Molecular Simulation of Lignocellulosic Biomass Components

Loukas Petridis, Center for Molecular Biophysics, Oak Ridge National Laboratory, Oak Ridge, TN

Lignocellulosic biomass comprises the vast majority of biomass on Earth and has the potential to play a major role in generation of renewable biofuels if cost-effective conversion can be achieved. Largely composed of plant cell walls, lignocellulosic biomass is a complex biological composite material that shows significant recalcitrance towards the structural deconstruction and enzymatic hydrolysis into sugars that is necessary for fermentation to bioethanol. Here, we report on studies combining molecular dynamics simulation and small angle neutron scattering to better understand structural properties of biomass components cellulose and lignin. Lignin, a major polymeric component of plant cell walls, forms aggregates in vivo after pretreatment of lignocellulosic biomass for ethanol production. The aggregates are thought to reduce ethanol yields by inhibiting enzymatic hydrolysis of cellulose. Extensive atomistic molecular dynamics simulations were used to demonstrate that the surfaces of the aggregates are highly irregular and characterized by a surface fractal dimension that is invariant across length scales from ~1–1000Å. The simulations reveal extensive water penetration of the aggregates and heterogeneous chain dynamics corresponding to a rigid core with a fluid surface. The detailed multiscale structure revealed here should aid in understanding biomass recalcitrance to hydrolysis and in feedstock engineering efforts to improve biofuel yield.

See more of this Session: Advances In Biofuels: DOE Bioenergy Research Centers I
See more of this Group/Topical: Sustainable Engineering Forum

(308e) Surface Patterning for Selective Bioconjugation of Oxide-Based Biosensors

Bradley W. Biggs[1], Heather K. Hunt[1] and Andrea M. Armani[2], (1)Mork Family Department of Chemical Engineering and Materials Science, University of Southern California, Los Angeles, CA, (2)Mork Family Department of Chemical Engineering and Materials Science & Ming Hsieh Department of Electrical Engineering-Electrophysics, University of Southern California, Los Angeles, CA

Thefabrication of on-chip, oxide-based biosensors that exhibit both highsensitivity and specificity is of significant interestto fields like security, food safety evaluation, and environmental monitoring,where rapid and real-time detection of pathogens is necessary. Surface functionalization strategies, whichimpart specificity towards a biochemical species of interest through thetethering of a recognition element to the sensor surface, play an importantrole in the development and final utility of these devices. Typicalsurface functionalization chemistries, such as the coupling of recognitionelements to silicon / silica sensor surfaces via silanecoupling agents, often result in the functionalization of the entire platformsurface, rather than the selective functionalization of the device. To overcome this, passivationstrategies consisting of combinations of grafting reactions and self-assemblytechniques are often used to selectively "pattern" the final devicesurface. However, this process can becumbersome, and detrimental to device performance, especially for on-chipoptical biosensors, such as the silica-on-silicon microtoroidalwhispering gallery mode resonators (Fig. 1), whose performance can benegatively impacted by wet chemistry techniques. Here, we present an alternative process forthe selective functionalization of silica biosensors via a simple dry etchingtechnique during device fabrication.This dry etching technique uses XeF_2 vapor, instead of themore typical KOH wet etch, to remove silicon, resulting in silica microtoroids supported by silicon pillars on a siliconsubstrate. Using this method, we canselectively tailor the silica surface with recognition elements, such asbiotin, which can be used to detect (strept)avidin, without alsofunctionalizing the surrounding silicon platform.

Todemonstrate the selectivity of this process, we created a variety of samplesusing the typical fabrication and surface functionalization techniques thathave been used to tailor the surface of ultra-high sensitivity microtoroidal whispering gallery mode resonators[1]. The fabrication of silica microtoroidsfrom a 2 micron thermal oxide on (100) silicon wafers consists of three simplesteps: (1) patterning circular pads of silica using typical photolithography,(2) undercutting the silica pads using silicon etching techniques to formsilica disks supported by silicon pillars, and (3) reflowing the silica disksto form silica microtoroids. From there, the biosensors are functionalizedthrough a three step functionalization process, thatin this example tethers a biotin recognition element to the surface: (1) O_2plasma treatment, (2) chemical vapor deposition of organosilanecoupling agents, and (3) attachment of the biotin recognition element to thecoupling agent via grafting. Here, weexamined the impacts of dry vs. wet etching (to remove silicon from thesurface) on the resulting ability to selectively functionalize the surface ofthe silica vs. the surface of the silicon.After the silicon etch, Atomic Force Microscopy(AFM) measurements were taken to determine surface roughness. X-ray Photoelectron Spectroscopy (XPS) was takenat various points throughout the fabrication and functionalization process toquantify the surface chemistry. Lastly,Fluorescence Microscopy (FM) measurements of the functionalized surfaces wereperformed to determine the relative coverage of the surface functionalizationon silica vs. silicon resulting from the different etchtypes.

As expected, the samples which wereetched using KOH did not allow for selective functionalization of the silicasurface on a silicon/silica wafer.Additionally, the XPS results showed that the oxygen plasma treatmentwas unable to remove residual potassium residue on the surface. In contrast, the samples etched using XeF_2were pristine after the oxygen plasma, with no elements from the etching stepdetected on the XPS, even at trace levels.Subsequent fluorescent microscopy studies further verified that it ispossible to selectively functionalize only the silica surface withoutadditional chemical patterning.

The abilityto leverage the inherent chemical properties of the substrate to encouragebinding only on the sensor surface will improve the sensor's overallperformance and collection efficiency.This opens the door for the creation of more complex on-chip structuresin not only for biosensors, but also for many other multi-materialdevices.

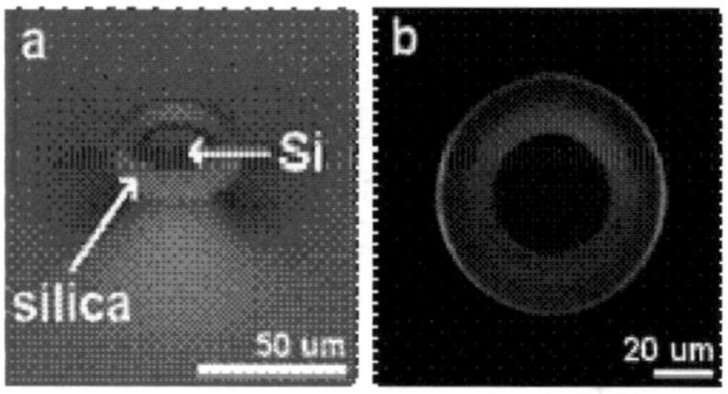

Figure 1. (a) Silica microtoroidaloptical resonators on silicon pillars; (b) Biotin-functionalized silica microtoroids labeled with Texas Red-Avidinfluorescent dye.[1]

1. H. K.Hunt, C. Soteropulos, and A. M. Armani, "Bioconjugation Strategies forMicrotoroidal Optical Resonators," Sensors. **10**, 9317-9336 (2010).

See more of this Session: Micro and Nanofabricated Sensors II
See more of this Group/Topical: Topical 9: Sensors

(13b) Mixed Rare Earth Oxides (REOs) for Desulfurization and Tar Removal From Gasifier Effluents

Rui Li, Joseph Bridges, Weishi Kong and Kerry.M. Dooley, Chemical Engineering, Louisiana State University, Baton Rouge, LA

The removal of H_2S, NH_3 and tars from syngas generated in biomass or coal/biomass gasification plays an important role in further processing of syngas. We have developed materials capable of both H_2S adsorption and tar reforming at high temperatures (~600-650°C). These are mixed rare earth oxides (REOs) such as CeO_2/La_2O_3, prepared by templated sol-gel methods, and then impregnated with transition metals (Mn, Fe, etc.). Desulfurization tests have been carried out using model syngas mixtures. Successive desulfurization-regeneration cycles some of these materials (e.g., $Mn/Ce/LaO_x$) demonstrated reasonable sulfur capacities in the presence of water and CO_2, along with regenerability. Naphthalene and propane have been used as models for tar reforming/cracking, and the effects of H_2S, CO and CO_2 on the model reactions were examined. H_2S and CO_2 are strong inhibitors, while CO is not. Fresh, used and regenerated sorbents and catalysts have been characterized by BET, TPR, XRD, XANES and EXAFS.

See more of this Session: Catalytic Conversion of Renewable Resources to Synthesis Gases and Pyrolysis Oils
See more of this Group/Topical: Sustainable Engineering Forum

(491f) The Binding Properties of Cellulases On Allomorphs Cellulose and Pretreated Biomass During Enzymatic Hydrolysis

Dahai Gao, **Shishir Chundawat**, Venkatesh Balan and Bruce Dale, Chemical Engineering and Material Science, Michigan State University, Lansing, MI

The depolymerization of cellulose to monomeric glucose is the key step for cellulosic ethanol biorefinery process. Cellulases binding on biomass is the essential first step for effective breakdown of biomass to sugars. A new fast flow liquid chromatography (FPLC) based method has been developed to quantify endoglucanase I (EG I, GH family 7B), cellobiohydrolase I (CBH I, GH family 7A) and cellobiohydrolase II (CBH II, GH family 6A) individually in the heterogeneous reaction environment.

The binding level of individual enzymes during 48 hours hydrolysis was studied on different cellulose allomorphs: micro crystal cellulose Avicel (cellulose Iβ), liquid ammonia treated cellulose (cellulose III), sodium hydroxide treated cellulose (cellulose II) and phosphoric acid swollen amorphous cellulose (AC). The digestibility rank is AC>cellulose III>cellulose II>cellulose I. However, the AC has the highest initial binding capacity while cellulose III has the lowest. After the substrate solubilized, most of the cellulase returned to supernatant. But CBH II is less stable. Only around 60% CBH II can be recovered.

The time course binding studies were also performed under pretreated biomass. Ammonia Fiber Expansion (AFEX) treated corn stover (CS), dilute acid treated CS and ionic liquid pretreated CS are compared. The results show the existence of lignin is responsible for a significant amount of unproductive binding. For AFEX-CS, the initial binding level for those cellulases are >80%. However, after 48 hours digestion, only around 50% of CBH I, 28% of CBH II and 19% of EG I are detected in the supernatant. The remaining cellulases are believed to be bound unproductively onto lignin. Similar results were found on ACID-CS. The results are critical for improving our understanding of enzyme synergism, productive/unproductive enzyme binding and the role of pretreatment on enzyme accessibility to lignocellulosic plant cell walls, as well as help to engineer novel low unproductive binding enzymes and develop an economic enzymes recycle system.

See more of this Session: Advances In Biofuels: DOE Bioenergy Research Centers I
See more of this Group/Topical: Sustainable Engineering Forum

(491c) Rheological Properties of Biomass Deconstruction Process Using Ionic Liquid

Alejandro G. Cruz[1], Jeff Mentel[2], Seema Singh[3] and Blake Simmons[3], (1)Deconstruction, Joint BioEnergy Institute, Emeryville, CA, (2)Malvern Instruments LTD, MA, (3)Joint BioEnergy Institute, Emeryville, CA

Lignocellulosic biomass must be pretreated prior to enzymatic hydrolysis to improve sugar yields and saccharification efficiency. Ionic liquid (IL) pretreatment has shown great potential as a novel pretreatment technology with sugar yields as high as ~100%. However, to improve process economics, higher biomass loading is necessary. Biomass slurry in IL are highly viscous. To decrease production cost, the biomass slurry should be pumpable and exhibit least resistance to flow. Here we investigate rheological properties of neat IL, biomass slurry in ionic liquid through dissolution and regeneration processes. Complex viscosities and modulus (both elastic and viscous) measurements provide a detail insight of biomass solubilization and cellulose regeneration process and pinpoint the onset of cellulose regeneration event. By generating small-angle oscillatory frequency sweeps of pretreated switchgrass at various increased loading percentages we can characterize the nature of the slurry where the material has strong intermolecular association. For a viscous material such as in the 3% pretreated switchgrass sample, the viscous modulus is dominant over the elastic modulus. Just a slight increase in percent loading makes the slurry behave like a gel like structure and the elastic modulus is dominant over the viscous modulus. All percent loading pretreated switchgrass show frequency dependence and shear thinning behavior.

See more of this Session: Advances In Biofuels: DOE Bioenergy Research Centers I
See more of this Group/Topical: Sustainable Engineering Forum

(10f) Effect of AFEX Pretreatment Degradation Products On Enzymatic Hydrolysis and Microbial Fermentation by Saccharomyces Cerevisiae 424A (LNH-ST)

Xiaoyu Tang[1], Leonardo da Costa Sousa[1], Shishir Chundawat[1], Mingjie Jin[1], Lau Ming[1], James Humpula[1], Nirmal Uppugundla[1], Kevin Chambliss[2], Ramin Vismeh[3], A. Daniel Jones[3], Zeyi Xiao[4], Bruce Dale[1] and **Venkatesh Balan**[1], (1)Chemical Engineering and Material Science, Michigan State Univeristy, Lansing, MI, (2)Chemistry and Biochemistry, Baylor University, Waco, TX, (3) Biochemistry and Molecular Biology, Michigan State University, East Lansing, MI, (4)School of Chemical Engineering, Sichuan University, Chengdu, CA, China

Several degradation products are formed or released during AFEX pretreatment process, which can be divided into five groups: 1) carbohydrates, 2) aliphatic acids, 3) nitrogenous compounds, 4) furans, 5) aromatic compounds. Some of these degradation products are expected to be inhibitory to *Saccharomyces cerevisiae* 424A(LNH-ST), thus affecting the xylose fermentation, as well as various saccharolytic enzymes. In this work, the effect of various AFEX degradation products on co-fermentation by *Saccharomyces cerevisiae* 424A(LNH-ST) were tested individually and in combinations. A synthetic media composed of similar nutrient composition as AFEX corn stover hydrolyzate was designed and successfully used as a control fermentation media in this study. Nitrogenous compounds showed the inhibitory effect on xylose fermentation, followed by carbohydrates, aliphatic acid and aromatic compounds. The inhibitory effect caused by nitrogenous compounds and carbohydrates were attributed to the presence of amides and oligomeric xylan, respectively. To our knowledge, this is the first report of inhibitory effect of oligomeric xylan and amides on xylose fermentation by *Saccharomyces cerevisiae*. When testing the amides (feruloyl amide, coumaroyl amide and acetamide) and their corresponding acids (ferulic acid, coumaric acid and acetic acid) individually, the amides were less inhibitory than their acids with respect to cell growth and xylose fermentation. From this point, the shift of degradation products from organic acids to amides may be responsible for the high fermentability of AFEX pretreated lignocellulosic biomass. We also evaluated the extent of enzymatic inhibition by various molecular fractions of AFEX treated corn stover degradation products. The pretreated biomass extracted products were fractionated using ultracentrifugation, gel filtration, and C18 resin solid phase extraction and further analyzed by mass spectrometry. The results indicate that both large and small molecule compounds play a role in enzyme/microbial inhibition.

See more of this Session: Biological Conversions and Processes for Renewable Feedstocks I
See more of this Group/Topical: Sustainable Engineering Forum

(491e) Microfluidic Assays for Biofuels Research

Rajiv Bharadwaj[1], Aarthi Chandrasekaran[2], Chieh Chang[3], April Wong[4], Blake Simmons[2], Paul D. Adams[2] and Anup Singh[2], (1)Sandia National Laboratories, Joint BioEnergy Institute, Livermore, CA, (2)Joint BioEnergy Institute, Emeryville, CA, (3)Sandia National Laboratories, Joint BioEnergy Institute, Livermore, (4)Sandia National Laboratories, Livermore, CA

We are developing next-generation assays for biofuels research using microfluidic technologies to provide significant improvement over conventional platforms in throughput, sensitivity, multiplexing and speed of analysis. To develop these microscale assays we utilize technologies covering a wide spectrum of complexity based on the requirements of the assays. For example, using off-the-shelf components we have developed a microtiter plate-based assay for screening cellulases against heterogeneous substrates. This approach relies on in-situ biomass

regeneration in micro-volumes to volumetrically meter biomass (sub-mg loading) and also precisely control the amount of residual IL for screening activities of novel IL-tolerant cellulases. We have also developed more sophisticated platforms for screening of lignocellulolytic enzymes using microfluidic chip for electrophoretic analysis of glycans. These chips are being used to screen genetically-engineered enzymes, assess performance of pretreatment processes, and for discovering cellulase activities in microbial communities. We are also developing high throughput microchips to perform combinatorial DNA assembly for optimization of biofuel synthesis pathways.

See more of this Session: Advances In Biofuels: DOE Bioenergy Research Centers I
See more of this Group/Topical: Sustainable Engineering Forum

(305b) Lifecycle Assessment and Policy: Implications of the Renewable Fuel Standard for Upper-Midwest Energy Supply

Michael J. Brodeur-Campbell[1], Jordan Klinger[1], Kathleen Halvorsen[2] and David Shonnard[1], (1)Chemical Engineering, Michigan Technological University, Houghton, MI, (2)School of Forest Resourses and Environmental Science and Department of Social Sciences, Michigan Technological University, Houghton, MI

LifecycleAssessment and Policy: Implications of the Renewable Fuel Standard for

Upper-MidwestEnergy Supply

Michael Brodeur-Campbell[1], Jordan Klinger[1],Kathleen Halvorsen[2,3], and David Shonnard[1,4]

[1] Department of ChemicalEngineering, Michigan Technological University

[2] Department of SocialSciences, Michigan Technological University

[3] School of Forest Resourcesand Environmental Science, Michigan Technological University

[4] Director Sustainable FuturesInstitute, Michigan Technological University

The use of biofuels derived from lignocellulosic biomass has receivedsignificant public attention with Congressional policies such as the EnergyPolicy Act of 2005 (EPAct), updated in the EnergyIndependence and Security Act of 2007 (EISA) which included the Renewable FuelStandard (RFS). The RFS mandates theblending of ethanol and advanced / cellulosic biofuels into the U.S.transportation fuel supply for each year until 2022. "The Renewable Fuel Standard has a lifecycleperspective and contains lifecycle greenhouse gas emissions reductionrequirements (relative to petroleum gasoline) for the production ofbiofuels." Conventional ethanol derivedfrom corn starch must meet a 20% reduction level and is capped at 15 billiongallons/yr. "°Cellulosic" ethanol mustmeet a 60% reduction standard and is expected to make up 16 billion gallons/yr.by 2022. "Advanced" biofuel that meet a50% reduction must make up the remaining5 billion gallons/yr mandated by 2022 (36 billion gallons/yr total ethanol). In addition to demanding an accounting for bothdirect and indirect land use change CO_2 emissions, the policy alsospecifies several other environmental and social metrics to consider inevaluating biofuel production.Environmental metrics specified include airquality, water quantity and quality, wetlands and ecosystem health, andwildlife habitat. Other indirect effectson society include energy security, commercial fuel production andinfrastructure, consumer fuel prices, job creation, agriculture impacts, ruraleconomic development, and future food prices.

While many studies have been performed on agricultural residues such ascorn stover (Sheehan, Aden et al. 2004, Hsu 2010) and wheat straw (Kabel, Bos et al. 2007), muchless attention has been given to woody feedstocks. For this LCA, several regionally importantpotential woody feedstocks are analyzed (hybrid poplar, hybrid willow, andmixed hardwood logging residues) as well as two herbaceous feedstocks (switchgrassmonoculture, and a diverse prairie grass ecosystem). This LCA is directly driven by therequirements of the RFS and includes metrics for water and air quality, as wellas global warming potential. Eutrophicationpotential was chosen as the most important measure for water quality, as theeffects of fertilizer use in biomass cultivation generally dominate allcategories of water quality.

Particulatematter emissions were chosen as the most important measure for air quality, asthis category has the greatest implication for human health.

SimaPro 7.2 and the EcoInventdatabase were used to model the process and assign environmental burdens. A co-product credit is assigned to therenewable electricity generated from the non-fermentable portion of thebiomass. This electricity is modeled todisplace grid electricity at the regional mix for the states of Michigan,Wisconsin, Minnesota, Iowa, and Illinois, which is derived approximately 70%from coal-fired power plants. Inputs andoutputs for the conversion were based on the National Renewable EnergyLaboratory Technical Report NREL/TP-510-32438 (Aden, Ruth, et al. 2002). The ratio of fuel to electricity producedfrom each feedstock was modified based on typical values for cellulose,hemicellulose, lignin, ash, and extractives for each feedstock. Table 1 below summarizes the results forGlobal Warming Potential, Eutrophication Potential, and Particulate MatterEmissions for the feedstocks analyzed.

Table1: Summary of LCA Results for Cellulosic Ethanol

	Logging Residue	**Prairie**	**Switchgrass**	**Willow**	**Poplar**
Global Warming Potential (g CO_2 eq. / MJ)	9.0	-11.0	0.9	-2.4	-7.9
Eutrophication Potential (g N eq. / MJ)	0.0	0.0	0.3	0.1	0.1
Particulate Matter <2.5 m (mg / MJ)	0.4	-1.4	-1.1	-2.8	-3.1

All fuel pathways analyzed meet 60% greenhousegas reduction requirements for a cellulosic biofuel, even before any co-productcredit is applied. With co-productcredit some pathways result in net negative carbon emissions due to coaldisplacement. Eutrophication potentialis increased in all cases due to feedstock production activities; the highestis for switchgrass which is the most fertilizer-intensive feedstock, while thelowest is for multispecies prairie. Particulateemissions are reduced for all cases except logging residues, largely due toavoided coal power. Particulate matteremissions for logging residues are high because of the low productivity peracre and long transportation distances required to supply enough material forconversion.

Future work will focus on refining ourfeedstock conversion emission estimates using the ASPEN Plus process simulationsoftware, and on including land use change estimates for carbon flows.

References

Aden, A., M. Ruth, K. Ibsen, J. Jechura, K. Neeves, J. Sheehan,B. Wallace, L. Montague, A. Slayton, and J. Lukas (2002). *Lignocellulosic Biomass to Ethanol ProcessDesign and Economics Utilizing Co-Current Dilute Acid Prehydrolysisand Enzymatic Hydrolysis for Corn Stover.* National Renewable EnergyLaboratory, Golden, Colorado.

Hsu,D., D. Inman, G.A. Heath, E.J. Wolfrum, M.K. Mann,and A. Aden (2010). *Life CycleEnvironmental Impacts of Selected U.S. Ethanol Production and Use Pathways in2022.* Environmental Science &Technology; 44, 5289-5297.

Kabel, M.A., Bos,G., Zeevalking, J., Voragen,A.G.J., Schols, H.A., (2007). *Effect of pretreatment severity on xylansolubility and enzymatic breakdown of the remaining cellulose from wheat straw.*Bioresource Technology; 98, 20342042.

Sheehan J, Aden A, Paustian K, Killian K,Brenner J, Walsh M, et al. (2004) *Energyand environmental aspects of using corn stover for fuel ethanol*. Journal of Industrial Ecology;7(4):117e46.

See more of this Session: Life Cycle Analysis of Renewable Feedstock-Based Processes and Products I
See more of this Group/Topical: Sustainable Engineering Forum

(305a) Life Cycle Assessment of Advanced Regional Biomass Processing Depots Integrated with Sustainable Cellulosic Feedstock Landscapes

Pragnya L. Eranki, Bryan Bals, Seungdo Kim and Bruce E. Dale, Department of Chemical Engineering, Michigan State University, Lansing, MI

Lignocellulosic biomass can help fulfill escalating demands for liquid fuels and mitigate the environmental impacts of petroleum-derived fuels. Cellulosic biofuels also offer potential solutions to many socio-economic and sustainability problems caused by petroleum dependence. However, the large biorefineries which process cellulosic biomass to fuels face many logistical problems because they are fully integrated, centralized facilities in which all units of the conversion process are present in a single location. Biomass logistical problems are further exacerbated by physical characteristics of feedstocks such as low bulk density, compositional variability, seasonality and their tendency to decompose.

A potential solution to these problems might be a network of distributed processing facilities called “Regional Biomass Processing Depots” (RBPDs). In their simplest configuration these depots procure, preprocess /pretreat, densify and deliver feedstock to the biorefinery while providing a single co-product animal feed to end-users. However, the RBPDs can be configured into other, more complex setups by employing multiple technologies such as the ammonia fiber expansion pretreatment (AFEX), Leaf Protein Concentrate (LPC) extraction, leaf/stem separation, pyrolysis and anaerobic digestion, as dictated by feedstock characteristics and local conditions, to produce a variety of valuable co-products and for energy generation via synergies among these technologies.

In a previous analysis we considered the differences between distributed processing and centralized processing combined with a variety of landscape scenarios. This study established that combining high yielding perennial grasses with distributed processing that includes densification of AFEX treated biomass prior to transportation to the biorefinery may generate greater energy yields and lower greenhouse gas emissions compared to centralized processing. In this study we aim to extend this analysis to watershed- scales in various landscape configuration scenarios using more sophisticated geospatial information system tools as well as by including a wider range of feedstocks and sustainable management practices, including the use of marginal lands. On the processing side, we evaluate the energy balances and environmental impacts of advanced RBPDs and their associated transport systems. We then intend to integrate the various elements within this overall system and conduct a life cycle analysis (LCA) using the Gabi software.

Through an integrated systems-wide analysis and by developing flexible models we seek the most sustainable landscape-processing-transportation arrangements for cellulosic biofuel production. We believe that greater energy and environmental benefits can be added to biofuel production by not merely aiming at improved logistics but by reimagining agricultural systems and landscapes, minimizing the challenges associated with biomass supply and its processing and by investigating multiple, potentially synergistic processing technologies and related complexities.

See more of this Session: Life Cycle Analysis of Renewable Feedstock-Based Processes and Products I
See more of this Group/Topical: Sustainable Engineering Forum

(11d) Determination of Binding Interactions Using Optical Microcavities

Carol Soteropulos, Department of Biomedical Engineering, University of Southern California, Los Angeles, CA, Heather K. Hunt, Mork Family Department of Chemical Engineering and Materials Science, University of Southern California, Los Angeles, CA and Andrea M. Armani, Mork Family Department of Chemical Engineering and Materials Science & Ming Hsieh Department of Electrical Engineering-Electrophysics, University of Southern California, Los Angeles, CA

Determinationof Binding Interactions Using Optical Microcavities

Carol E. Soteropulos[1, 2],Heather K. Hunt[2, 3], Andrea M. Armani[1,3,4]

[1]*Department of Biomedical Engineering, University of Southern California, Los Angeles, California 90089, USA*

[2]*Department of Biological Engineering, University of Missouri,Columbia, Missouri, 65211, USA*

[3]*Mork FamilyDepartment of Chemical Engineering and Materials Science, University ofSouthern*

California, Los Angeles, California 90089, USA

[4]*Ming Hsieh Departmentof Electrical Engineering-Electrophysics, University of Southern California,*

Los Angeles, California 90089, USA

Opticaldetection techniques, ranging from fluorescent-based assays to label-freesurface plasmon resonance (SPR) sensors, enable researchers not only to detecttrace substances and perform diagnostics, but also to probe the fundamentalnature of numerous biological systems.For example, SPR sensors are routinely used to determinesubstrate-enzyme binding kinetics. Thedetermination of accurate binding kinetics for molecular systems is fundamentalto the understanding of the interactions between biomolecules within a bindingpair. For example, the way in whichindividual protein molecules of an enzyme interact (association / dissociationequilibria, molecular positioning, etc.) within enzyme-substrate complexes isthe key to their specificity and ability to catalyze a reaction. This information is subsequently used todesign improved and more targeted therapeutics.

Anemerging label-free optical detection technique is the silica opticalmicroresonator, which, while originally designed for telecommunications, hasshown great applicability for detection and diagnostics. These evanescent-field devices have very lowoptical losses, and are therefore able to confine light for long periods oftime at specific resonant wavelengths. Whena molecule binds to the surface of the cavity, the effective refractive index ofthe optical field is modified, and the resonant wavelength changes, thusenabling the sensing capabilities of these devices. The overall sensitivity (detection limit) ofthis platform is determined by the optical loss or the photon lifetime of themicrocavity. Due to their low opticalloss, microcavity devices have demonstrated very high sensitivities towards avariety of biological and chemical targets, including single molecules, singleviruses, and single nanoparticles.[3-5]

Althoughmuch work has been done demonstrating the sensing and diagnostic capabilitiesof these devices, they have not been explored for their applicability as probesfor kinetics measurements. One of theprimary hurdles in using optical microcavities to study binding kinetics is thedevelopment of a robust and stable surface chemistry that can immobilize onehalf of the binding pair without degrading the optical performance of thedevice. A covalent attachment isnecessary to determine binding kinetics because the probe molecule must bereliably attached to the surface in order to obtain consistent informationregarding the probe-ligand system.Covalent attachment of probe molecules using silanecoupling agents, in conjunction with vapor deposition techniques, has beenshown to address these requirements for on-chip optical resonators, such as themicrotoroid.[6-7] Here, we adapt these facile protocols togenerate silica microsphere optical resonators that present with a uniformsurface coverage of biotin probe molecules.This method ensures that the microsphere resonators can be used ashighly sensitive (strept)avidin sensors, due to the high biotin / (strept)avidin bindingaffinity. Specifically, we modify theprotocols to better maintain the optical sensitivity of microsphere resonantcavities. We demonstrate the maintenanceof optical performance of a functionalized spherical resonant cavity beforeexamining the detection capabilities towards streptavidin of the functionalizedoptical cavities in aqueous solution and analysis of the binding kinetics.

Thesurface chemistry (uniformity, biological activity) was verified usingfluorescent microscopy. Before and afterfunctionalization, the optical performance of the microcavity was characterizedto determine the impact of the chemistry on the device. The photon lifetime or quality factor (Q) ofthe device was > 1 million after probe attachment, well above the Q-thresholdnecessary to detect single viruses or particles. Finally, the site stability was determinedafter long-term storage (>60 days).After characterizing the surface functionalization, the functionalizeddevices were used to detect 1 nM streptavidin in anaqueous solution. Subsequently, wedetermined the binding kinetics constants based on the resonant wavelength shiftduring detection, using a reversible bimolecular reaction model for the bindingpair to calculate the dissociation constant (k_d) through a simple exponential functiondescribing the dissociation phase [8].

Our measured constants agreed with those previously determined throughother methods for the immobilized biotin-streptavidin system. We also showed the limiting impact of masstransport on the determination of the association constant using such devices. Similar mass transport limitations have beencommented upon by many researchers in the surface plasmon resonance community[8].

Asa result of the covalent attachment of biotin to the microsphere surface, weare able not only to detect low concentrations of biotin, but also to determinebinding kinetics of the biotin-streptavidin pair. Through this type of highly sensitive analysis,we can better quantify the interaction between two molecules, potentially atthe single molecule level, and gain important information regarding interactionstrengths, as well as the specific way in which molecules interact during achemical reaction. In order to calculatean association constant (k_a), we would need to take into account anumber of separate issues which arise as a result of one molecule of the pairbeing immobilized on a surface, such as steric hindrance and mass transportlimitations [9]. This demonstrationmarks the first example of the use of such novel optical platforms for probingfundamental biological questions.

References

1. H.K. Hunt, and A. M. Armani, "Label-Free Biological and ChemicalSensors," Nanoscale **2**,1544-1559 (2010).

2. X.D. Fan, I. M. White, S. I. Shopoua, H. Y. Zhu, J. D. Suter, and Y. Z. Sun,"Sensitive optical biosensors for unlabeled targets: A review,"Analytica Chimica Acta **620**, 8-26(2008).

3. A.M. Armani, R. P. Kulkarni, S. E. Fraser, R. C. Flagan, and K. J. Vahala,"Label-Free, Single-Molecule Detection with Optical Microcavities,"Science **317**, 783 (2007). (publishedonline July 5, 2007 [DOI: 10.1126/science.1145002].).

4. F.Vollmer, S. Arnold, and D. Keng, "Single virus detection from the reactiveshift of a whispering gallery mode," Proceedings of the National Academyof Sciences of the United States of America **105** (2008).

5. J.Zhu, S. K. Ozdemir, Y.-F. Xiao, L. Li, L. He, D.-R. Chen, and L. Yang,"On-chip single nanoparticle detection and sizing by mode splitting in anultrahigh-Q microresonator," Nature Photonics **4**, 46-49 (2009).

6. H.K. Hunt, and A. M. Armani, "Recycling Microcavity OpticalBiosensors," Optics Letters **36**(2011).

7. H.K. Hunt, C. Soteropulos, and A. M. Armani, "Bioconjugation to OpticalMicroresonators," Sensors **10**,9317-9336 (2010).

8.S. Zhao, and W. M. Reichert, "Influence of Biotin Lipid SurfaceDensity and Accessibility of Avidin Binding to the Tip of an Optical FiberSensor," Langmuir **8**, 2785-2791(1992).

9.P. Schuck, and A. P. Minton, "Analysis of Mass Transport-LimitedBinding Kinetics in Evanescent Wave Biosensors," Analytical Biochemistry **240**, 262-272 (1996).

See more of this Session: Biosensor Devices I
See more of this Group/Topical: Topical 9: Sensors

(62d) Medium Optimization for Enhanced Lipid Production by Chlorella Protothecoides UTEX25

Kuan-Chen Cheng[1], Ming Ren[1] and Kimberly Ogden[2], (1)Chemical and Environmental Engineering, The University of Arizona, Tucson, AZ, (2)University of Arizona, Tucson, AZ

Algal biodiesel as the third generation biofuels has received considerable attention in recent years because of its renewable, biodegradable, and nontoxic fuel properties. Algal also exhibit several advantages over other energy crops, such as higher photosynthetic efficiency, higher biomass production and lipid content, and faster growth rate. The composition of algal-produced hydrocarbons vary among the microalgae species resulting in many different lipids, hydrocarbons, and other complex oils.

In the present study, a fresh water microalgae, *Chlorella protothecoides* (UTEX25), was evaluated for its lipid productivity through medium optimization. A one-factor-at-a-time approach was applied to determine the influence of each ingredient of proteose medium on both biomass and lipid production. $MgSO_4$°E7H_2O, $NaNO_3$, $CaCl_2$°E2H_2O, K_2HPO_4 and KH_2PO_4, and NaCl were removed individually from the culture medium, and the results were compared with the one obtained from the original medium. After 11-day cultivation, the results demonstrated that the removal of MgSO4°E7H_2O has positive effect on both biomass growth and lipid production. The biomass reached 0.95 g/L (the biomass production obtained from original medium was 0.54 g/L). The lipid content also increased to from 4.6 to 9.5 % when MgSO4 was removed. On the other hand, the removal of $NaNO_3$ and proteose adversely reduced biomass growth. The biomass production decreased from 0.54 to 0.37 and 0.29 g/L, respectively, when $NaNO_3$ and proteose was removed. Therefore, MgSO4°E7H_2O, $NaNO_3$, and proteose were chosen for further evaluation.

A statistical experimental design, response surface methodology (RSM), was applied to determine the optimal concentration of the three nutrients for both biomass and lipid production. The predicted results of these three nutrients are validated in comparison with experimental data. The change of the lipid profiles obtained from original and optimized medium are evaluated using GC-FID. Results are applied to minimize nutrient requirements in large-scale pond cultivation.

See more of this Session: Advances In Algal Biorefineries I
See more of this Group/Topical: Sustainable Engineering Forum

(254d) The Impurities Removal In Polysilicon

Guoqiang Huang and Qiuling Shi, Department of Chemical Engineering, Tianjin University, Tianjin, China

Polysilicon, as a base material, is widely used in the electronic information and photovoltaic industry. The Japan earthquake, happened on March 11, and the subsequent nuclear leakage caused a rise in polysilicon price. An American market research organization reported that about 25% of the global silicon was suspended due to the earthquake. At the same time, nuclear leakage made it understandable that solar energy has found favor in clean energy. Our work aims at the purification of polysilicon in Modified Simens Process and the methods adopted to remove boron and phosphorus impurities affect the product quality most, which are important problems hamper the industry's development.

At present, Modified Simens Process which accounts for about 85% of the total output is the best way to produce polysilicon. The feedstock trichlorosilane(TCS) used here requires that the purity is above 99.999%, the content of boron and phosphorus elements is no more than 10ppb, polymers 10ppm, total metal elements is less than 10ppb. It is of great importance to control the content of impurities. Here 26 metallic impurities, 5 non-metallic impurities, polysilanes, organosilane compounds et al. were summarized.

New ways to removal boron and phosphorus impurities contain: 1) Reactive distillation; 2) Distillation – Absorption; 3) Wet nitrogen bubble method; 4) Adsorptive distillation. Our work is to exploit new purification technology, integrate the existing technological process and improve purification efficiency. Energy saving and environmental protection are also discussed. Reactive distillation and adsorptive distillation are two methods we use.

The key thing in reactive distillation is the choice of reactant. Reactant suitable for industrial application should contain these features: 1) It will react with boron or phosphorus compounds to form stable complex, quickly and efficiently. 2) The boiling point of it and its complex must be higher than the components of chlorosilanes, only in this way will the excess reactant and the complex formed be removed easily. 3) It will not react with the component of chlorosilanes, unless there is a special need. 4) It never reacts with components of the reactor, vessel, column and their accessory devices to introduce new impurities into the chlorosilane system. 5) The price and amount should be suitable for large-scale production. Some kinds of reactants are recommended.

Accordingly, choosing the adsorbents in adsorptive distillation, we should regard physical and chemical properties of the components and impurities in the system, e.g. boiling point, polarity. The adsorbent should be of the following characters. 1) It should have bigger adsorption amounts and adsorb impurities effectively at appropriate temperature and pressure. 2) It will not release impurities into the system.3) It has a uniform particle size and large surface area. 4) Regeneration is easy. Activated charcoal, silica gel and some special adsorbents are good choices.

We also proposed two technological processes using the two methods mentioned above to reduce the boron and phosphorus in polysilicon. The simulation result is good.

See more of this Session: Process Development and Design for Solar-Grade Silicon Production
See more of this Group/Topical: International Congress on Energy 2011

(305c) Optimal Design and Operations of Cellulosic Biofuel Supply Chains Under Uncertainty

Fengqi You, Dept of Chemical Engineering, Carnegie Mellon University, Pittsburgh, PA and Belinda Wang, Department of Chemical and Biological Engineering, Northwestern University, Evanston, IL

Concerns about climate change, energy security, and the diminishing supply of fossil fuels are causing our society to search for new renewable sources of transportation fuels. Domestically available biomass has been proposed as part of the solution to our dependence on fossil fuels. Biofuels, especially liquid transportation fuels produced from cellulosic materials, have the benefits of significantly reducing greenhouse gas (GHG) emissions and leading to new jobs and greater economic vitality in rural areas. Renewable Fuels Standard (RFS), part of the Energy Independence and Security Act of 2007, establishes a target of 16 billion gallons of cellulosic biofuel annual production by 2022 [1, 2], although the United States produced less than 1 billion gallons of liquid fuels from cellulosic materials in 2010 [3, 4]. In observance of this mandatory production target, many new cellulosic biomass-to-biofuels supply chains will be designed and developed in the coming decade for better economic, environmental and social performances. However, uncertainty resulting from supply and demand variations may have significant impact on the biofuel supply chain. Therefore, an efficient optimization strategy is urgently needed to for the design and operations of sustainable and robust cellulosic biofuel supply chains.

In this work, we address the optimal design and planning of cellulosic biofuel supply chains under supply and demand uncertainty. A two-stage stochastic mixed-integer linear programming (SMILP) model combined with Monte Carlo sampling and the associated statistical analysis [4, 5] is proposed to deal with different types of uncertainty, and it is incorporated into a multi-period planning model that takes into account the main characteristics of the biofuel supply chains, such as seasonality of feedstock supply, biomass deterioration with time, geographical diversity and availability of biomass resources, feedstock density, diverse conversion technologies and byproducts, infrastructure compatibility, demand distribution, regional economic structure, and government incentives. In the two-stage framework, the supply chain network design and capacity planning decisions are made "here-and-now" prior to the resolution of uncertainty, while the production, transportation and storage decisions for each time period are postponed in a "wait-and-see" mode. The SMILP model integrates decision making across multiple temporal and spatial scales and simultaneously predicts the optimal network design, facility location, technology selection, capital investment, production operations, inventory control, and logistics management decisions. In order to solve the resulting large scale SMILP problems effectively, a decomposition algorithm based on sampling average approximation [5] and multi-cut L-shaped method [6, 7] is proposed by taking advantage of the problem structure.

In addition to the economic objective of minimizing the annualized net present cost, the SMILP model is also extended to integrate with life cycle assessment (LCA) and regional economic input-output (EIO) analysis through a multiobjective optimization scheme to include two other objectives: the environmental objective measured by life-cycle greenhouse gas emissions and the social objective measured by the number of accrued local jobs resulting from the construction and operation of the cellulosic biofuel supply chain. The multiobjective optimization framework allows the model to establish tradeoffs among the economic, environmental, and social performances of the cellulosic biofuel supply chains in a systematic way. The multiobjective optimization problem is solved with an ε-constraint method and produces Pareto-optimal curves that reveal how the optimal annualized cost and the supply chain network structure change with different environmental and social performance of the entire supply chain [10, 11].

The proposed optimization model and solution method is illustrated through county-level case study for the state of Illinois. Three major types of biomass, including crop residues, energy crops, and wood residues, and three major conversion pathways, including biochemical conversion, gasification followed by Fischer-Tropsch synthesis and fast pyrolysis followed by hydroprocessing are considered. Uncertainty information is generated from the time series analysis [8] based on the historical data of biomass feedstock supply [9] and liquid fuel demand [1]. County-level results will be presented that provide regionally-based insight into transition pathways of biomass production and conversion. Computational results also demonstrate the effectiveness of the proposed decomposition algorithm for the solution of large-scale SMILP problems.

References:

[1] Biomass Program Multi-Year Program Plan 2010; EERE, U.S. DOE, March 2010.

[2] National Biofuels Action Plan; Biomass Research and Development Board: U.S. U.S. Energy Information Administration.

[3] National Renewable Fuel Standard Program for 2010 and Beyond; U.S. EPA, February 2010

[4] Department of Agriculture and U.S. Department of Energy: 2008.

[5] Shapiro, A.; Homem-de-Mello, T., A simulation-based approach to two-stage stochastic programming with recourse, Mathematical Programming, 1998, 81, 301-325

[6] Birge, J.R.; Louveaux, F., Introduction to Stochastic Programming, Springer Verlag, New York, 1997

[7] You, F.; Wassick, J. M.; Grossmann, I. E., Risk management for global supply chain planning under uncertainty: models and algorithms. AIChE Journal 2009, 55, 931-946.

[8] Enders, W., Applied Econometric Time Series. Wiley: Hoboken, NJ, 2004.

[9] National Agricultural Statistics Service.

[10] You, F.; Tao, L.; Graziano, D. J.; Snyder, S. W., Optimal Design of Sustainable Cellulosic Biofuel Supply Chains: Multi-objective Optimization Coupled with Life Cycle Assessment and Input-Output Analysis. AIChE Journal 2011, In press, DOI: 10.1002/aic.12637.

[11] You, F.; Wang, B., Life Cycle Optimization of Biomass-to-Liquids Supply Chains with Distributed-Centralized Processing Networks. Industrial & Engineering Chemistry Research 2011, Submitted.

See more of this Session: Life Cycle Analysis of Renewable Feedstock-Based Processes and Products I
See more of this Group/Topical: Sustainable Engineering Forum

(76h) Portable Surface Plasmon Resonance Biosensor for Detecting Marine Toxin Domoic Acid

Steve L. Bonilla, Gar S. Tsun and **Qiuming Yu**, Chemical Engineering, University of Washington, Seattle, WA

Abstract

Domoic acid (DA) is a neurotoxin produced by marine diatoms of the genus *Pseudo-nitzschia* that can accumulate in edible shellfish. Amnesic shellfish poisoning (ASP) is a serious illness that is caused by eating shellfish contaminated with DA and in some cases it can be fatal. The increasing intensity and frequency of toxic blooms has drawn more attention to develop fast and effective techniques to detect DA. High-performance liquid chromatography with ultraviolet detection (HPLC-UVD) is the current official method for detecting DA. Unfortunately, due to the complexity and relatively large size of the equipment, HPLC-UVD detection needs to be performed by trained personnel in the laboratory. Surface plasmon resonance (SPR) biosensors have been demonstrated to provide a fast and sensitive DA detection. However, the size of conventional SPR biosensors makes them unsuitable for working in the field. Newer developed surface plasmon resonance coupler and disperser (SPRCD) biosensor are significantly smaller and lighter than conventional SPR biosensors and could potentially become the perfect tool for detecting DA on-site.

In this paper we present the detection of DA using a SPRCD biosensor and compare the detection sensitivity with a conventional spectroscopic SPR biosensor. The inhibition assay was used for both SPRCD and SPR biosensors with polyclonal anti-DA antibody. A salt calibration was performed in order to be able to directly compare the response of the SPRCD biosensor to that of the conventional SPR biosensor. DA was immobilized onto the gold coated surface of the sensor chip previously functionalized with a mixed self assembled monolayer (SAM) of $HS(CH_2)_{11}(C_2H_5O)_4OH$ and $HS(CH_2)_{11}(C_2H_5O)_6NH_2$ thiols. The OH-terminated thiol serves as a protein non-fouling background while the NH2-terminated thiol provides a functional group to which DA molecules could bind via the NHS/EDC chemistry. The ratio of OH to NH_2 terminated thiols was optimized for both types of sensors in order to reduce the non-specific binding but in the meantime to provide sufficient functional groups for immobilization of DA. Polyclonal anti-DA and DA were mixed to the final concentrations of 1 - 5 μg/mL for polyclonal anti-DA and 0.1 – 1500 ng/mL for DA prior to flowing through the sensors. The detection curves were obtained for both sensors and, finally, the limit of detection and the working range were determined. The influence of surface fictionalization on the sensitivity and specificity of detection as well as SPRCD sensing mechanisms will be discussed.

(10c) Exploration of Significant Enzyme Activities Required for the Deconstruction of Ionic-Liquid Pretreated Biomass

Christopher J. Barr[1], Jeffery Mertens[2] and Constance Schall[1], (1)Chemical and Environmental Engineering, University of Toledo, Toledo, OH, (2)NCAUR, USDA-ARS, Peoria, IL

Terrestrial biomass is a second generation renewable feedstock, comprised of polysaccharides derived from cellulose and hemicellulose, which can be used to produce carbon-based chemicals and fuels. The recalcitrance to deconstruction of biomass is one of the main impediments in the conversion of biomass to monomeric sugars and the production of sugar platform fuels and chemicals. Ionic Liquid (IL) pretreatment creates unique biomass substrates composed of amorphous cellulose with largely intact hemicellulose and lignin. IL pretreated substrates differ from many other pretreatments which remove hemicellulose and retain cellulose crystallinity. This substrate poses an interesting challenge for enzymatic deconstruction using traditional commercial mixtures.

Mono-component cellulase and hemicellulase enzymes from *Aspergillus nidulans* have been expressed, purified, and characterized. Key enzyme activities and/or interactions that significantly contribute to the conversion of a variety of lignocellulosic biomass substrates to monomeric sugars were examined by enzyme combinations in series of hydrolysis experiments. Auxiliary hemicellulases seem to be essential to the mono-component enzyme mixtures. Investigation of these accessory enzymes also yields insight into an understanding of the differences in hemicellulose structure between feedstocks. Some auxiliary hemicellulases include acetyl xylan esterases (AXE), arabinofuranosidases, and α-glucuronidases. In addition to mono-component enzyme mixtures, next generation commercially-available enzyme mixtures were characterized and formulated using key activities to optimize hydrolysis at low enzyme loadings. These simplified mixtures will be dependent on the substrate. As part of these studies, an analysis of variance (ANOVA) was used to statistically model the main and interaction effects of key enzyme activities for each substrate.

See more of this Session: Biological Conversions and Processes for Renewable Feedstocks I
See more of this Group/Topical: Sustainable Engineering Forum

(76c) Microfluidic-Based Sensor for Safety Application

Joseph Parisi Jr., Ying Wang and Yu Lei, Chemical, Materials, and Biomolecular Engineering, University of Connecticut, Storrs, CT

Microfluidic-basedsensor for safety application

Joseph Parisi,Ying Wang, Yu Lei

Departmentof Chemical, Materials, & Biomolecular Engineering, University ofConnecticut, 191 Auditorium Road, Unit 3222, Storrs, Connecticut 06269

Food andwater safety and the detection of food- and water-born pathogens is still amajor concern for the food and water industry. Current techniques lack sensitivity,versatility, and quick detection. Research in the area of microfluidics hasalready solved some of these problems by creating various separation anddetection devices. However, many of these devices require the use of separateequipment to prepare the sample for testing. Very little research has beenperformed on combining these devices to create a complete detection devicecapable of working without the need of macro scale equipment. In this study, we will integrate a sample separator with a simple detector for real timeseparation and detection of targets. The proof-of-concept will be demonstratedusing a model particle and the analytical characteristics will be investigated.Its application for real pathogen detection will be explored.

See more of this Session: Biosensor Devices II
See more of this Group/Topical: Topical 9: Sensors

(76g) Biosensor Applications for Polymeric Nanosensors

Kevin J. Cash and Heather A. Clark, Pharmaceutical Sciences, Northeastern University, Boston, MA

Recent years have shown the incredible versatility and utility of polymeric nanosensors. Sensors have been developed for a wide variety of ionic analytes (*e.g.* sodium, potassium, pH) as well as nonionic analytics (*e.g.* glucose) and have been applied for challenging research fields such as detection of sodium sparks in live cardiomyocytes. This field of research has yielded new knowledge on cellular metabolism and function. Additionally, these sensors can be applied *in vivo*, through implantation in or under the skin in animals; showing potential as a research tool for continuous monitoring of analyte concentrations without the need for direct samples to be taken from the animal.

To date the choice of recognition element has limited the range of targets which can be measured with these polymeric nanosensors. This stems from the need for small molecule recognition elements (e.g. ionophores) which can be incorporated into the polymer matrix. Through the combination of these nanosensors with enzymatic recognition elements we can now detect a far wider variety of biologically relevant targets through detection of metabolic byproducts of enzymatic reactions.

This work was supported by DARPA and the National Institute of General Medicine of the National Institutes of Health under award number R01 GM084366.

See more of this Session: Biosensor Devices II
See more of this Group/Topical: Topical 9: Sensors

(13d) Catalytic Cracking of Oak Pyrolytic Vapors Via Bench-Scale In-Situ Fixed-Bed Catalysis

David J. Mihalcik, Akwasi A. Boateng, Charles A. Mullen and Neil Goldberg, Eastern Regional Research Center, Agricultural Research Service, USDA, Wyndmoor, PA

Pyrolysis oils derived from lignocellulosic feedstocks by way of fast pyrolysis offer the possibility of a distinct alternative to the use of petroleum-based fuels. However, an inherent drawback associated with pyrolysis oil its high oxygen content which is due to the nature of the lignocellulosic biomass from which it derives. Extensive efforts are being put forth to treat pyrolytic vapors in situ using zeolite catalysts. For our experiment, the catalytic conversion of oak pyrolysis vapors was performed at 500°C using a 2.5 kg/hr fluidized-bed fast pyrolysis reactor that was fitted with a re-heated (425°C) split-stream fixed-bed catalytic cracker containing zeolite-type catalyst. The placement of the split-stream fixed-bed catalytic reactor was varied within the condenser train to determine the effect of the removal of increasing amounts of water and oxygenated components from within the pyrolytic vapor stream. The cracked pyrolytic vapors resulting from the fixed-bed catalyst were collected by way of immediate contact with a dry-ice acetone bath followed by secondary collection using a methanol spray condenser. Karl Fisher titration and elemental and quantitative GC/MS analyses were performed to determine relevant trends within the experiment. The recovered liquid products that passed through the fixed-bed catalytic reactor displayed a remarkable decrease in the amount of oxygen containing species and a significant increase in the presence of aromatic hydrocarbons.

See more of this Session: Catalytic Conversion of Renewable Resources to Synthesis Gases and Pyrolysis Oils
See more of this Group/Topical: Sustainable Engineering Forum

A Novel Reactive Distillation Approach for Producing 1,3-Dihydroxyacetone from Glycerol

Xi Hong, Carl T. Lira, and Dennis J. Miller
Department of Chemical Engineering and Materials Science, Michigan State University
East Lansing, Michigan 48824

Introduction

Glycerol is a viable renewable feedstock that has emerged as a consequence of triglyceride conversion to biodiesel. In addition to glycerol conversion to commodity chemicals such as propylene glycol and epichlorohydrin, formation of higher-value specialty chemical products such as 1,3-dihydroxyacetone (DHA) will provide income for the biorefinery.

DHA is the active ingredient in sunless tanning products. Currently, DHA is industrially produced by fermentation of glycerol, but this microbial approach is challenging because of low production rates and high recovery costs [1]. As a result, the high selling price of DHA limits access of sunless tanning products in the general population.

Recently, scientists have attempted conventional chemical oxidation processes to make DHA from glycerol using metal catalysts with air or molecular oxygen. The goal is to achieve direct oxidization of the secondary hydroxyl group in glycerol to a carbonyl bond. Unfortunately, this reaction is always accompanied with primary alcohol oxidation and a variety of products (i.e. glyceroaldehyde, glyceric acid, etc.) are produced. The highest yield obtained from this method is in the range of 40-60% [2-4].

We present here a novel chemical process to convert glycerol to DHA that greatly improves the efficiency and reduces cost. The essential feature of this process is the selective formation of a 6-membered acetal of glycerol via reactive distillation, followed by selective oxidation of the secondary alcohol to its corresponding ketone product using ruthenium metal and a chemical oxidant. The ketone product is then hydrolyzed to obtain DHA.

A continuous process of making DHA from glycerol has been designed and will be presented. Reactions involved in the process are rapid and can be carried out under mild conditions. Yields of individual process steps are all in excess of 90%, suggesting that this process has good potential to reduce DHA production costs.

Experiments

Materials

Anhydrous glycerol (99.5%), 1-propanol (99.9%) and 1-butanol (99.9%) were obtained from J.T. Baker. Ethanol (99.5%), 1,1-diethoxyethane (99%) and sodium hypochlorite (15% aqueous solution) were purchased from Pharmco and Aldrich, respectively. Sodium carbonate (> 99%), sodium bromate (> 99%), potassium bromate (> 99.8%), calcium hypochlorite (> 98%), sodium periodate (> 99%), sodium iodide (> 99%) were purchased from Sigma-Aldrich. Amberlyst-15 cationic exchange resin (16-50 mesh), purchased from Sigma-Aldrich, was investigated as the catalyst. Before being used in reactions, it was pretreated at 90°C in a vacuum oven for 2 hours. Ruthenium dioxide hydrate (99.9% trace metals basis, 28% water content) and ruthenium trichloride hydrate (reagent plus, 46% Ru content) were obtained from Aldrich. Hydrogen (99.999%) and helium (99.999%) were used for GC analysis; both were purchased from Linde Gas, Inc.

Analysis

Gas Chromatography Analyses

1-Butanol and 1-propanol were used as internal standard and solvent, respectively, for this analysis. To prepare samples for analysis, 0.16 ml 1-butanol and 0.2 ml of reaction sample were added to a volumetric flask and the mixture was diluted with 1-propanol to 10 ml of solution. A Perkin-Elmer Sigma 2000 gas chromatograph with FID and a DB-Wax capillary column (15m × 0.53mm × 1.0 μm) was used for analysis. Ultra high purity helium was used as the carrier gas; the column temperature was ramped from 40°C (3min) at a rate of 30°C/min to 220°C and held for 1 min. The total program time was 10 min. On column injection was applied to analyze the samples. Injector and detector chamber temperatures were set at 60°C and 300°C, respectively.

GC-MS analysis

The structure of some products was identified using GC-MS analysis (GC-Agilent 6890, MS-Agilent 5973). A 1μL sample was separated in a DB-wax capillary column (30m × 0.25mm × 0.25 μm). The initial column temperature was 40 °C (for 5min), the temperature was then increased at 20 °C/min to 250 °C and held for 10 min. Total program time was 24.5 min. Injector temperature was set at 240 °C. Electron impact (EI) mass spectra were monitored from 20 to 300 m/z. The MS source was set at 230 °C.

Results and discussion

Synthesis of glycerol acetal

GA isomers were synthesized at 45 °C by mixing glycerol (3.1 kg, 33.9 mol), DEE (4.0 kg, 33.9 mol), and Amberlyst-15 cationic exchange resin (120 g) and stirring for 4 h. Following reaction and removal of volatile species (DEE and ethanol) by rotary evaporation, GC-MS analysis of the product mixture showed the presence of four different structures corresponding [5, 6] to the four GA isomers displayed in Table 1.

Table 1. Structure and IUPAC name of GA isomers

	Structure	IUPAC Name	Designation
a)	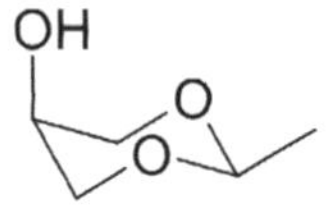	cis-5-hydroxy-2-methyl-1,3-dioxane	Cis-6
b)	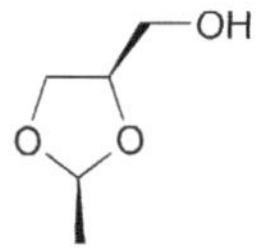	cis-4-hydroxymethyl-2-methyl-1,3-dioxolane	Cis-5
c)	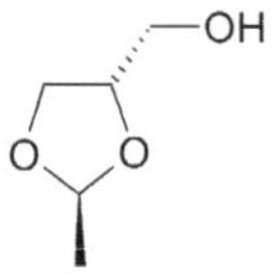	trans-4-hydroxymethyl-2-methyl-1,3-dioxolane	Trans-5
d)		trans-5-hydroxy-2-methyl-1,3-dioxane	Trans-6

GA isomerization equilibrium

In the presence of acid catalyst [7, 8], the four GA isomers interconvert and reach an equilibrated composition. To demonstrate this equilibration, three samples obtained from the distillation experiments below with widely differing isomer compositions (cis-6 from 0% to 84%, cis-5 from 2% to 54%, trans-5 from 2% to 41%, and trans-6 from 0% to 63%) were stirred for 5 hr at 25 °C in the presence of 5 wt% (relative to GA) Amberlyst-15 cationic exchange resin. Identical isomer compositions were observed in the products from the three samples: cis-6, cis-5, trans-5 and trans-6 were 37±1%, 17±1%, 13±1%, and 33±1%, respectively. The molar percentages of 6-membered isomers (cis-6 + trans-6) and 5-membered isomers (cis-6 + trans-5) are approximately 70% and 30%, respectively, at equilibrium.

Rapid isomerization requires vigorous agitation (to ensure mass transfer between solid catalyst and solution) and elevated temperature. For example, Sample 1 took 12 h to reach equilibrium at 0 °C, but only one hour at 80 °C. Increasing temperature also shifts the equilibrium composition. For Sample 1, increasing temperature from 0 °C to 80 °C decreased the sum of cis-6 + trans-6 isomers from 78% to 54%. However, the total quantity of the cis- form isomers (cis-6 + cis-5) and trans- form isomers (trans-6 + trans-5) isomers remained essentially constant over all temperatures at 55±2% and 45±2%, respectively.

Production of the cis-6 GA isomer

A glass vacuum distillation column was constructed to separate the four GA isomers. The column (height = 3.6 m, i.d. = 50 mm) was packed with 0.6 cm 316 Stainless Steel PRO-PAK distillation packing (Scientific Development Co.). The column reboiler, consisting of a two-liter round-bottom flask, was heated with a 750 W electric heating mantle. Vapor temperature was measured at the top of the column.

In the initial distillation, 896 g (800 ml) of a GA mixture containing 35% cis-6, 18% cis-5, 15% trans-5, and 32% trans-6 isomers was placed in the reboiler. To avoid isomer interconversion, the solution was maintained at a weakly basic state by adding 50 g of sodium carbonate to the mixture. This experiment was carried out at a pressure of 20.5 ± 1.0 mm Hg (except for Sample 1, where the pressure was temporarily reduced to 9.0 mmHg).

Twelve 60-mL distillate samples were collected over the course of the distillation and analyzed by GC. Ninety percent of the initial charge to the still pot was recovered as distillate in these samples (Table 2). The material remaining in the still pot consisted primarily of the trans-6 isomer and glycerol; inclusion of this material gave essentially complete (100±2%) closure of the material balance for the experiment.

Table 2. Summary of GA isomers recovered in the initial distillation

	Initial GA charge		Product recovery in distillate		Percent recovered in distillate (%)
	Mass (g)	Composition (wt%)	Mass (g)	Percent of total	
Cis-6	311	35	288	36	93
Cis-5	166	18	152	19	92
Trans-5	132	15	120	15	91
Trans-6	287	32	246	30	86
Total	896	100	806	100	90

Figure 1 gives the reboiler liquid temperature, vapor temperature at the top of the column, and composition (via GC) of the distillate samples taken during the distillation. The results clearly show the order of decreasing volatility of the GA isomers as cis-6, cis-5, trans-5, and finally trans-6. As the more volatile isomers are removed from the still pot, liquid and vapor temperature gradually increase.

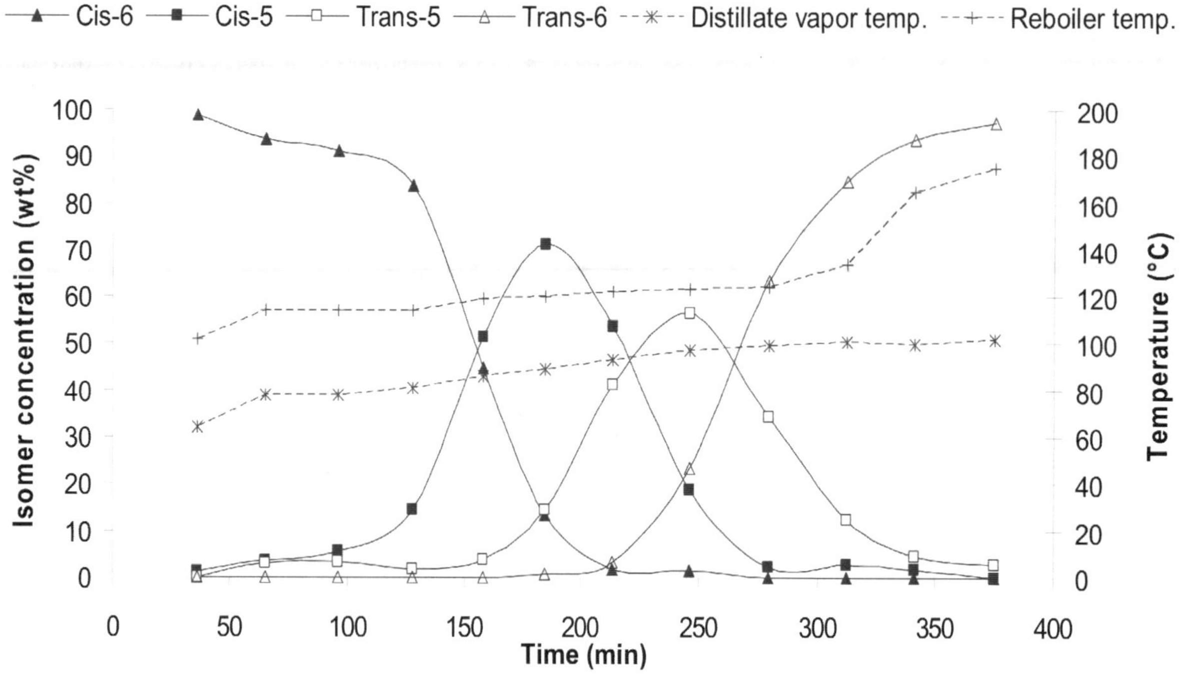

Figure 1. Initial distillation of GA isomer. Column pressure was 20.5±1.0 mmHg (except for the first sample, where the pressure was 9.0 mmHg).

Based on the result of the initial distillation and the fact that GA isomers can interconvert to each other and reach an equilibrated isomer composition in the presence of acid catalyst, a continuous reactive distillation experiment was then carried out to produce the cis-6 isomer as the sole product. In this run, 504 g (450 mL) of a mixture of 36 wt% cis-6, 16 wt% cis-5, 14 wt% trans-5, and 34 wt% trans-6 GA isomers was charged to the reboiler pot of the glass column along with 50 g of Amberlyst-15 cationic exchange resin. For this experiment, the column pressure was maintained at 2.9±0.2 mmHg. Once the distillate product started to condense at the top of the column, a continuous feed of the same GA isomer mixture was introduced to the column reboiler to maintain its liquid level constant. The feed rate was adjusted to match the distillate rate from the column, which was approximately 1.8 mL/min. During the distillation, 16 distillate samples were collected with an average volume of 59 mL.

As shown in Figure 2, this experiment produced samples with high concentration of cis-6 isomer (from 90% to 96%) continuously during operation of the column. The reboiler temperature and distillate vapor temperature were essentially constant during operation, indicating that the reboiler composition was constant and that the column was operating at steady state. The total quantity of distillate obtained was 1028 g, more than twice the amount of the initial charge to the reboiler, demonstrating that continuous production of the cis-6 isomer from the GA acetal mixture is feasible. Based on these results, it is fully expected that an appropriately designed commercial-scale column with sufficient separation capacity would provide essentially pure cis-6 isomer.

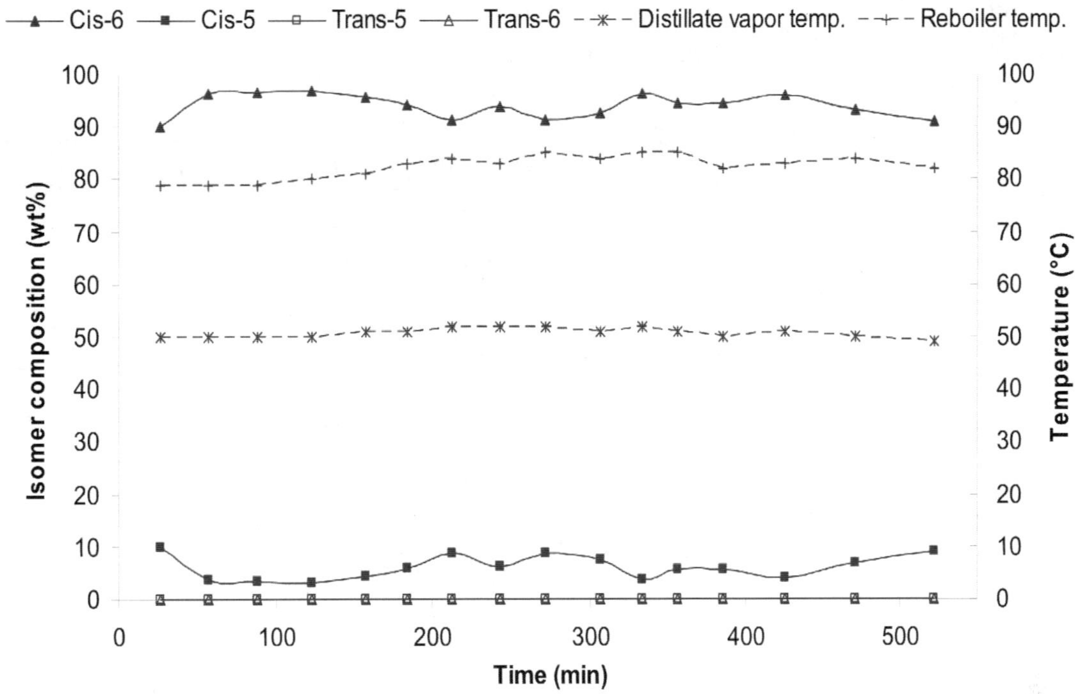

Figure 2. Continuous reactive distillation of GA isomers. Column pressure = 2.9±0.2 mm Hg.

Oxidation of cis-6 to the precursor of DHA

With the availability of cis-6 isomer, a new synthesis pathway can be developed to obtain valuable chemicals. For instance, by oxidizing the remaining free hydroxyl group in cis-6 and hydrolyzing the corresponding ketone product (2-methyl-1,3-dioxan-5-one, G-tone), 1,3-dihydroxyacetone (DHA) can be produced [7].

For this purpose, a variety of oxidation reagents including pyridinium chlorochromate (PCC), ruthenium tetroxide, hydrogen peroxide, molecular oxygen, and ozone were tested to convert cis-6 into its ketone product.

The most effective method is the ruthenium tetroxide-catalyzed oxidation. The reagents used in this reaction involves a ruthenium species (i.e. $RuCl_3$ and RuO_2) and another oxidizing salt (i.e. NaClO, $NaBrO_3$, $NaIO_4$, $Ca(ClO)_2$, $KBrO_3$, etc.) During reaction, the ruthenium species is first oxidized to RuO_4, and then RuO_4 oxidizes cis-6 to the corresponding ketone product. The co-product is a black precipitate of RuO_2, which can be re-oxidized to RuO_4 by the oxidizing salt in the aqueous solution. The regenerated RuO_4 then enters the organic phase again for another cycle of oxidation. In this way, the alcohol substrate can be continuously oxidized to the desired product with catalytic amounts of RuO_4 while the secondary oxidant is consumed. When the reaction is finished, the residue RuO_4 species in the organic phase can be easily recovered by adding isopropanol to make it RuO_2 precipitate [9].

Table 3 presents the results of cis-6 oxidation experiments catalyzed by ruthenium-based substances. It is clear that this type of reaction very efficient. In some reactions over 90% yield of G-tone was achieved within 30 min.

Table 3. Ruthenium tetroxide catalyzed oxidation of cis-6 at 25°C, 1.0 atm.

Reaction	Cis-6 (g)	Water (g)	Catalyst ($RuO_2 \cdot xH_2O$) loading (g)	Secondary oxidant solution/mass (g)/concentration	Time (min)	G-tone yield (%)
1	0.41	10	0	NaClO/4/10%	30	0
2	0.42	10	0	$KBrO_3$/10/saturated	1440	0
3	4.0	50	0.40	NaClO/100/10%	30	80
4	0.39	11	0.31	NaClO/4/10%	30	76
				NaClO/5/10%	60	95
5	0.45	10	0.05	$Ca(ClO)_2$/10/ saturated	20	87
6	0.40	10	0.30[a]	$NaIO_4$/10/saturated	75	88
7	0.42	10	0.26	$KBrO_3$/10/ saturated	45	> 99
8	0.43	10	0.32[a]	$KBrO_3$/10/saturated	75	95
9	4.1	10	0.40	$KBrO_3$/100/saturated	20	97
10	0.30	10	0.41	$NaBrO_3$/10/10%	20	96
11	10.0	10	0.51	$NaBrO_3$/100/20%	40	90
12	3.0	10	0.10	$NaBrO_3$/12.5/20%	20	95

a. $RuCl_3 \cdot xH_2O$ was used as catalyst.

Hydrolysis of G-tone to DHA

Hydrolysis of G-tone to DHA is very fast and can be carried out at mild conditions (Scheme 1). In one experiment, 5 wt% (weight to substrate) of Amberlyst-15 cationic exchange resin was added to the final mixture of Reaction 12 in Table 3. When the mixture was heated to 30°C, complete hydrolysis of G-tone was achieved in 10 min.

Scheme 1. Processes of catalytic conversion of glycerol into useful chemicals.

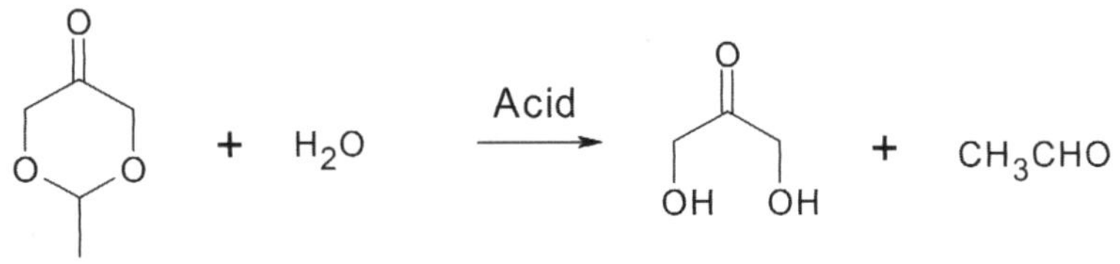

A novel process of making DHA from glycerol

Using cis-6 as an intermediate provides a relatively quick, environmentally safe and economical route to 1,3-dihydroxyacetone (DHA) from glycerol. Based on the experimental results obtained, a novel process of making DHA from glycerol is developed and described here.

Figure 3 presents the overall scheme of a novel DHA production process based on acetaldehyde or 1,1-diethoxyethane (DEE) as the glycerol acetal forming reagent. The process may also work with other aldehydes or ketones. Selection of the acetal forming reagent is dependent on price, availability and process feasibility.

In Figure 3, glycerol and acetaldehyde are introduced to Reactor 1 to form cyclic glycerol acetals. The reaction mixture, which contains four isomers of glycerol acetal, is then evaporated to get rid of water and unreacted acetaldehyde for recycling. After that, the mixture of glycerol acetal isomers is introduced to a vacuum distillation column, in which the cis-6 GA isomer is formed, distilled, and thus obtained from the top of the column. At the same time, other isomers, which have higher boiling points, react in the reboiler to re-establish isomerization equilibrium. The continuous recycling of the less-volatile glycerol acetal isomers through the equilibration-distillation procedure can transform all of the products into the desired 6-membered structure.

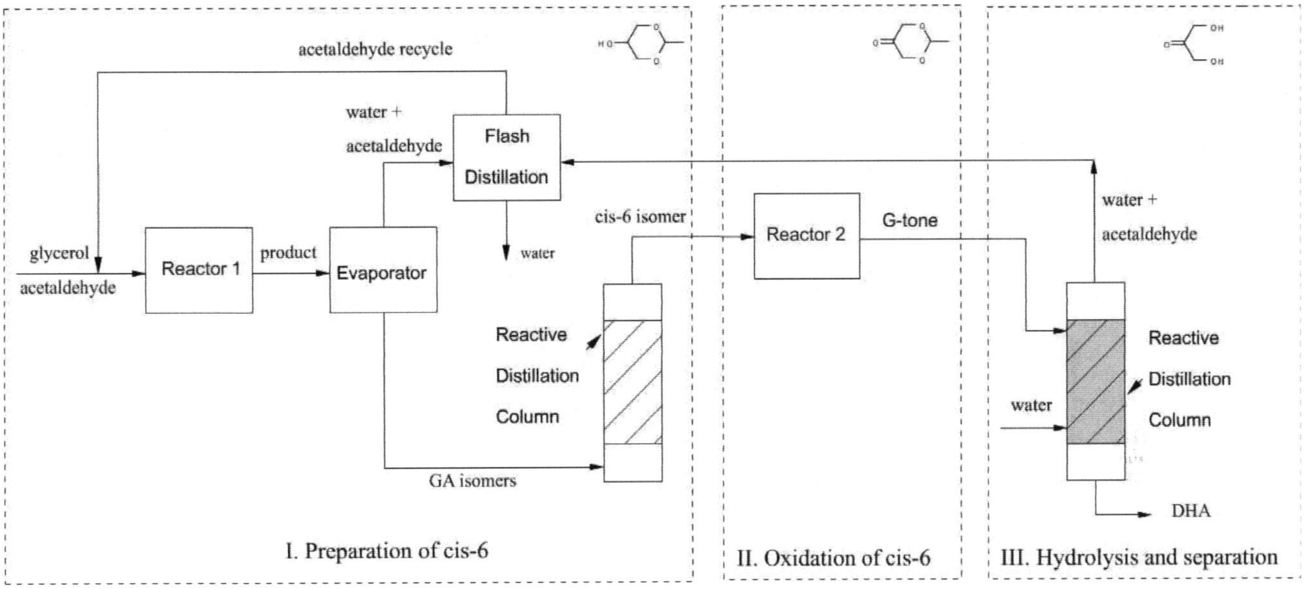

Figure 3. A novel process to produce DHA from glycerol.

In the second stage of this design, cis-6 GA isomer is introduced to the oxidation Reactor 2 to form 2-methyl-1,3-dioxan-5-one (G-tone). A preferred operation for this stage is to use ruthenium tetroxide-related oxidation.

In the last stage, G-tone and an excess amount of water are reacted, perhaps in a reactive-distillation column, to hydrolytically decompose G-tone to the desired product DHA and acetaldehyde, which can be recycled or reformed into DEE. The possible application of reactive distillation along with a large amount of water will overcome any chemical equilibrium limitations of this reaction, such that a high yield of DHA can be achieved from this process.

Summary

A novel process to make 1,3-dihydroxyacetone (DHA) from glycerol is presented in this paper. The process involves three major steps: 1) formation of glycerol acetal (GA) and reactive distillation to get the 6-membered GA isomer 5-hydroxyl-2-methyl-1,3-dioxane (cis-6); 2) oxidation of cis-6 GA isomer to the corresponding ketone product 2-methyl-1,3-dioxan-5-one (G-tone); 3) conversion of G-tone to DHA via hydrolysis. Experimental results show that in each step, over 90% yield of the desired product can be achieved. Thus, the new method holds a good potential for commercial success.

References

1. Mishra, R, Jain, S. and Kumar, A. (2008). Microbial production of dihydroxyacetone. Biotechnology Advances. 26, 293-303.
2. Garcia, R, Besson, M. and Gallezot, P. (1995). Chemoselective catalytic oxidation of glycerol with air on platinum metals. Applied Catalysis A: General. 127, 165-176.
3. Hu, W, Knight, D, Lowry, B. and Varma, A. (2010). Selective Oxidation of Glycerol to Dihydroxyacetone over Pt-Bi/C Catalyst: Optimization of Catalyst and Reaction Conditions. Industrial & Engineering Chemistry Research. 2010, 49, 10876–10882.
4. Kimura, H. (1993). Selective oxidation of glycerol on a platinum-bismuth catalyst by using a fixed bed reactor. Applied Catalysis A: General. 105, 147-158.
5. Da Silva Ferreira, A, Barbe, J. and Bertrand, A. (2002). Heterocyclic acetals f from glycerol and acetaldehyde in port wines: Evolution with aging. Journal of Agricultural and Food Chemistry. 50, 2560-2564.
6. Câmara, J, Marques, J, Alves, A. and Da Silva Ferreira, A. (2003). Heterocyclic acetals in Madeira wines. Analytical and Bioanalytical Chemistry. 375, 1221-1224.
7. Deutsch, J, Martin, A. and Lieske, H. (2007). Investigations on heterogeneously catalysed condensations of glycerol to cyclic acetals. Journal of Catalysis. 245, 428-435.
8. Aksnes, G, Albriktsen, P. and Juvvik, P. (1965). Studies of cyclic acetal and ketal isomers of glycerol. Acta Chemica Scandinavica. 19, 920-930.
9. Tojo, G. and Fernandez, M. (2006). Oxidation of alcohols to aldehydes and ketones. NY: Springer.

(308c) A Novel MEMS Preconcentration Approach for Analysis of Ketones and Aldehydes In Breath

Mingxiao Li[1], Xiao-an Fu[1], Souvik Biswas[2], Michael H. Nantz[2] and Richard M. Higashi[2], (1)Chemical Engineering, University of Louisville, Louisville, KY, (2)Chemistry, University of Louisville, Louisville, KY

A Novel MEMS PreconcentrationApproach for Analysis of Ketones and Aldehydes in Breath

Mingxiao Li[a], Souvik Biswas[b],Michael H. Nantz[b], Richard M. Higashi[b,c,d], Xiao-An Fu[*a],

Departments of [1]Chemical Engineering and [2]Chemistry,[3]Center for Regulatory and Environmental Analytical Metabolomics(CREAM), and [4] James Graham Brown Cancer Center,

The analysis of exhaled breath is a promising non-invasive diagnosticmethod for early detection of lung diseases. Certain volatile organic compounds(VOCs), such as ketones and aldehydes in exhaled breath, have already beenrecognized as metabolic indicators of many diseases including lung cancer andchronic obstructive pulmonary disease (COPD).

We report a MEMS-based micropreconcentratorthat has been engineered to selectively preconcentrateultra-trace gaseous ketones and aldehydesin exhaled breath. The silicon-glass micro-preconcentratorhas outer dimensions of 14mm14 mm, total inner surface area of ~300mm^2,and a total inner volume of ~7 L. The surface of the preconcentratoris functionalized with a quaternary ammonium aminooxycompound, for preconcentrating trace carbonyl VOCs bymeans of an oximation reaction. The concentrated ketone- and aldehyde- adducts aredirectly analyzed by Fourier transform-ion cyclotron resonance (FT-ICR) massspectrometer (MS). We demonstrate the utility and enhanced sensitivity of thisapproach in the detection of C1- to C12 ketones and aldehydes in exhaled breath.

See more of this Session: Micro and Nanofabricated Sensors II
See more of this Group/Topical: Topical 9: Sensors

(308b) Adapting Biomedical Nanoenabled Sensing to Environmental Sensing

Warren W. Layne, Superfund, U.S. EPA, Chicago, IL

Toward Environmental Nanoenabled Sensor Design Based On the DesignMethodology for Engineered Biomedical Sensors: A Cross Disciplinary Journey

This overview will highlight nanomaterialpreparations engineered for targeted drug delivery and in-vivo sensing that maybe conceptually useful in the preparation of nanomaterials for remediation andsensing of persistent environmental human and ecological toxins.

See more of this Session: Micro and Nanofabricated Sensors II
See more of this Group/Topical: Topical 9: Sensors

(62e) Single-Cell Growth Kinetics of Algae, Chlorela Vulgaris

Alim Dewan[1], Jihye Kim[2], Swastika S. Bithi[2], Marci Kerls[2], Siva A. Vanapalli[2] and M. Nazmul Karim[2], (1)Chemical Engineering, Texas Tech University, Lubbock, TX, (2)Chenical Engineering, Texas Tech University, Lubbock, TX

Single-cell growth kinetics of algae, *Chlorela vulgaris*

AlimDewan, JihyeKim, Swastika S. Bithi, Marci Kerls, Siva A.Vanapalli, M. Nazmul Karim*

Departmentof Chemical Engineering, Texas Tech University, Lubbock, Texas 79409, USA.

*E-mail: naz.karim@ttu.edu

Abstract

Traditionally,the growth kinetics of algae is investigated using cell culture equipment thatcontains large number of cells. The kinetic parameters are measured fromrepresentative samples of the cell culture and the measured parametersrepresent the statistical average of all the cells in the samples. In thisstudy, we have investigated the growth kinetics of single-cell of *Chlorellavulgaris* using droplet based PDMS microfluidic chip. The kinetic parameterswere determined by counting the total number of cells under investigationduring the entire period of the experiment. We found that there is significantheterogeneity in the growth and size of single cells and kinetic parameters aredifferent even if the initial cell numbers are the same. We found that thereis significant heterogeneity in the growth and size of single cells and kineticparameters are different even if the initial cell numbers are the same. Figure1exmplifies the heterogeneity of single-cell growth kinetics. We hypothesizethat the heterogeneity occurs due to the difference in initial cell conditionssuch as the age of the cell, and the size.

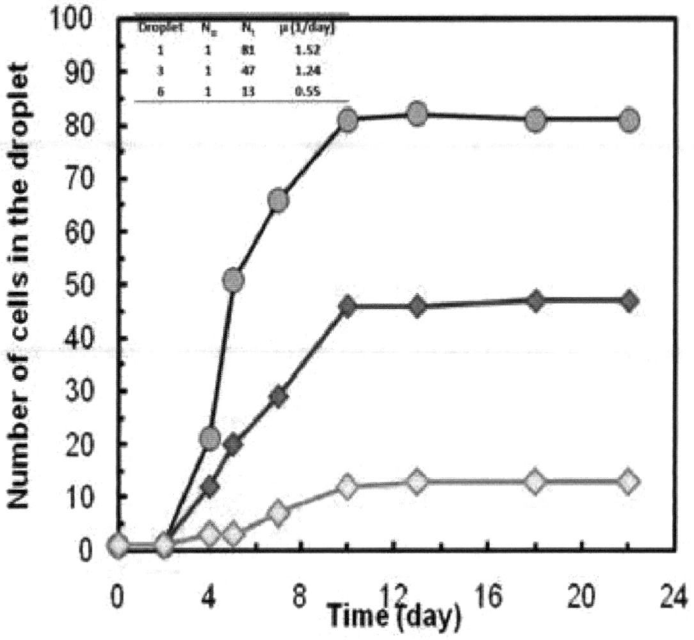

Figure1.Example of single-cell growth curves. Inset shows the initial cell number (N_o),final cell number (N_t) and the specific growth rate ().

See more of this Session: Advances In Algal Biorefineries I
See more of this Group/Topical: Sustainable Engineering Forum

(76a) Stimuli-Responsive Microfluidic Valves for Antigen Detection

Brad J. Berron and Chistopher N. Bowman, Department of Chemical and Biological Engineering, University of Colorado, Boulder, CO

There is increasing interest in the use of microfluidic devices in the detection and isolation of rare cellular species (circulating tumor cells and multipotent cells). These microfluidic capture approaches are typically limited to the recognition of a single surface marker on cells, and confirmation of the cellular identity is typically accomplished through immunofluorescence and microscopy. Here we demonstrate surface-mediated enzymatic polymerization as a simple approach to alter the flow pattern of a microfluidic device in response to the presence of a targeted biomaterial, and enable automated detection against additional biomarkers. In this novel approach, a flow-blocking polymerization event occurs only in response to the biodetection event of interest. When the antigen of interest is present, polymerization occurs rapidly, and prevents flow through the microchannel at pressures greater than 15 psig.

See more of this Session: Biosensor Devices II
See more of this Group/Topical: Topical 9: Sensors

(144e) Simultaneous Detection of Ascorbic Acid, Dopamine, and Uric Acid At Choline Monolayer Supported Multiwalled Carbon Nanotubes Film

Bing-Bing Xu and Hai-Jun Wang, Jiangnan University, Wuxi, China

Abstract: A novel, rapid and convenient method for the simultaneous detection of ascorbic acid (AA), dopamine (DA), and uric acid (UA) was developed at a multiwalled carbon nanotube (MWCNT)/choline (Ch) monolayer-modified glassy carbon electrode (GCE). The Ch was covalently immobilized on the surface of GCE through oxygen atom. The Ch monolayer provides a positively charged surface with $-N^{+}(CH_3)_3$ polar groups, so that it can attract negatively charged MWCNTs to the surface. Hence, the MWCNT/Ch film exhibited remarkable electrocatalytic activities towards the oxidation of AA, DA and UA due to the advantages of high electrode activity, large surface area, prominent antifouling property, and high electron transfer kinetics. Individual and simultaneous determination of AA, DA and UA were studied by all sorts of electrochemical means. All aforementioned analytes showed well-defined catalytic oxidation peaks at MWCNT/Ch/GCE. The peak separations between AA and DA, DA and UA are sufficiently large for their potential recognition and simultaneous detection in mixture. Under the optimum conditions, the designed MWCNT/Ch/GCE exhibited low detection limit, high sensitivity and wide linear range for simultaneous detection of AA, DA, and UA.

See more of this Session: Biosensor Devices III
See more of this Group/Topical: Topical 9: Sensors

(286f) A Sulfur-Sulfur Thermochemical Water Splitting Cycle for Thermal-to-Chemical Energy Conversion

Nick AuYeung, Malachi D. Bunn and Alexandre F.T. Yokochi, School of Chemical, Biological and Environmental Engineering, Oregon State University, Corvallis, OR

Thermochemical water splitting presents a way to store thermal energy in the chemical bonds of a H_2 molecule. Besides becoming an increasingly important energy carrier, hydrogen already has vast importance in petroleum upgrading and ammonia synthesis. In an effort to improve the Sulfur-Iodine Thermochemical Water Splitting Cycle, a novel Sulfur Sulfur Thermochemical Cycle may offer an easier, all fluid route to hydrogen than its predecessor:

Step 1A, liquid phase: $4I_2+4SO_2+8H_2O \leftrightarrow 4H_2SO4 + 8HI$

Step 1B, liquid phase: $8HI+H_2SO_4 \leftrightarrow H_2S+4H_2O+4I_2$

Step 2, gas phase: $3H_2SO_4 \leftrightarrow 3H_2O+3SO_2+1.5O_2$

Step 3, gas phase: $H_2S+2H_2O \leftrightarrow SO_2+3H_2$

An overall theoretical thermal efficiency assuming a maximum temperature of 1200 K is estimated to be 55%. Here, the I_2 consumed in Step 1A is conveniently regenerated in situ via Step 1B. Furthermore, H_2S easily desorbs from the liquid reaction medium and can then be steam reformed to produce H_2 and SO_2, (Step 3) the latter of which is recycled. The H_2S generation and steam reformation steps (Step 1 and Step 3 respectively) have been demonstrated in our laboratory.[1]

This presentation will focus on the parametric dependence of this highly endothermic, high temperature steam reformation reaction on reactant concentrations, temperature, and residence time in tubular reactor.

[1] Yokochi, A., AuYeung, N. In *A Proposed New Sulfur-Sulfur Thermochemical Cycle*, AIChE National Meeting, Nashville, Tennessee, Nashville, Tennessee, 2009.

See more of this Session: Concentrated Solar for Power Generation and Chemical Processing I

(11b) A New Design of Bio-Hybrid Nanopores for DNA/RNA Detection

Kwang Joo Kwak[1], Xuejin Wen[2], Wei-Ching Liao[3], Cherry Gupta[1], Bo Yu[4], Gintaras Valincius[5], David J. Vanderah[6] and James Lee[7], (1)Chemical and Biomolecular Engineering, The Ohio State University, Columbus, OH, (2)Electrical and Computer Engineering, The Ohio State University, Columbus, OH, (3)Mechanical and Aerospace Engineering, The Ohio State University, Columbus, OH, (4)Nanoscale Science and Engineering Center, Ohio State University, Columbus, OH, (5)Chemistry and Bioengineering, Institute of Biochemistry, Vilnius 08662, Lithuania, (6)Center for Advanced Research in Biotechnology, National Institute of Standards and Technology, Rockville, MD, (7)Chemical and Biomolecular Engineering, Ohio State University, Columbus, OH

A New Design of Bio-hybrid Nanoporesfor DNA/RNA Detection

Kwang Joo Kwak[1,2], Xuejin Wen[1,3],Wei-Ching Liao[1,4], Cherry Gupta[2], Bo Yu[1], GintarasValincius[5], David J. Vanderah[6], Wu Lu[1,3] andL. James Lee[1,2*]

[1]NSF Nanoscale Science and Engineering Center for Affordable Nanoengineering of Polymeric Biomedical Device (NSEC-CANPBD)

[2]Department of Chemical and BiomolecularEngineering

[3]Department of Electrical and ComputerEngineering

[4]Department of Mechanical Engineering

The Ohio State University, Columbus, Ohio 43210

[5]Institute of Biochemistry, Mokslininku 12,LT-08662 Vilnius, Lithuania

[6]Biomolecular Structure and Function Groupat the Center for Advanced Research in Biotechnology (CARB), National Instituteof Standards and Technology (NIST), Rockville, Maryland 20850

*Contact author-E-mail: lee.31@osu.edu

The dynamic nature of biomoleculetransport across the native cell membranes is difficult to understand due tothe complexity of the membranes and the transport phenomenon. Recent advancesin tethered biomimetic lipid membranes provide opportunities to study proteinadsorption, ion transport, and membrane mechanical properties. By selectingcell membrane lipids and adding proper protein, the tethered bilayer lipidmembrane (tBLM) with well-defined nanopore array can serve as a useful tool forthe investigation of various biological lipid structures and their interactionswith other biomolecules including cell membrane-DNA interactions.

The tBLM with well-defined nanopores wasdemonstrated on the SiO_2 surface with pre-drilled nanochannels. Thenanochannel array with from ~10,000 holes down to a single hole was prepared byE-beam lithography for electrochemical impedance spectroscopy (EIS) or electriccurrent measurements. To expose dimensions of tens of nanometers, a beamcurrent of 100 pA with an acceleration voltage of 100 kV was used. The spacingbetween two nanochannels in both x and y directions was 33 mm so that the number of nanochannelsis 10,000 with a circular testing area with the diameter of 4 mm. A thinself-assembled monolayer (SAM) was then formed on the Au layer. The tethered BLM was prepared by incubation with a diphytanoylphosphatidylcholine (DPhyPC) solution of 10 mMconcentration in ethanol for ~10 min and subsequently injection of PBS bufferwithin ~10 s. The atomic force microscopy (AFM) image with the nanopores on atBLM was observed in the liquid environment. The nanochannels with a diameterof from 100 nm down to 10 nm were drilled through the Au-coated $SiO_2$by e-beam lithography. Since the lipid bilayers could also form on thesidewalls of the nanowells, the nanopore size was smaller than the nanochannelsize as measured by AFM. The tBLMs with different pore size and pore densitywere used as DNA/RNA detection platforms.

The presence of nanochannels made thehigh frequency semicircle of the Cole-Cole EI spectrum incomplete, similar tothat observed when the densely-packed tBLM turned into a loosely-packed tBLM (lp)tBLM because of the reduced DPhyPC concentration. Unlike the (lp)tBLMs,this tBLM with well-defined nanopore array showed very stable and repeatable EIspectra. The Cole-Cole EIS spectra of SAMs and tBLMs with and without nanochannelswere measured and compared to the finite element method (FEM) simulation. Theresults of our current work show that both the morphology and dielectricproperties/conductivity of tBLM systems can be adjusted by varying experimentalparameters.

This bio-hybrid nanopore system with a-hemolysin is used to investigate the DNA-pore interactions by measuring both EIS and electric conductance whenlarge (e.g. vector pGFP) and small (e.g. microRNA) linear DNA/RNAs are used as model biomolecules.

See more of this Session: Biosensor Devices I
See more of this Group/Topical: Topical 9: Sensors

(10b) Combined Computational and Experimental Investigations of Processive Cellulases for Engineering Activity Improvements

Gregg T. Beckham[1], **Larry Taylor**[2], Michael F. Crowley[3], William S. Adney[3] and Michael E. Himmel[3], (1)National Bioenergy Center, National Renewable Energy Laboratory, Golden, CO, (2) Biosciences Center, NREL, Golden, CO, (3)Biosciences Center, National Renewable Energy Laboratory, Golden, CO

The Family 7 cellobiohydrolase (Cel7A) from *Hypocrea jecorina* (*Trichoderma reesei*) is one of the most well-characterized, active cellulases known for deconstruction of cellulose to soluble sugars. With the aim to engineer enhanced cellulase specific activity, we have investigated the Cel7A mechanism on cellulose with molecular dynamics simulations and biochemical experiments. We have elucidated new functions for the carbohydrate-binding module, determined the effect of glycosylation on the stability of the linker and carbohydrate-binding module, examined the role of aromatic and polar residues in the catalytic domain tunnel, and examined the structure of the catalytically-active complex of Cel7A on crystalline cellulose up to the microsecond time scale. These new insights combined with experimental activity data have led to a large-scale experimental screening of new enzymes to examine both natural diversity and engineered enzymes for activity and stability improvements. Our combined computational and experimental approach has demonstrated that unprecedented improvements in specific activities of processive cellulases are possible, which when combined in industrial cocktails, will result in lower enzyme loadings for cell wall deconstruction and lower overall production costs for renewable biofuels.

See more of this Session: Biological Conversions and Processes for Renewable Feedstocks I
See more of this Group/Topical: Sustainable Engineering Forum

(641a) Direct Conversion of Lignocellulosic Biomass to Drop-In Hydrocarbon Fuel

Xuejun Pan and Li Shuai, Biological Systems Engineering, University of Wisconisn, Madison, Madison, WI

This research reports an innovative process (HDA) for producing fuel-grade hydrocarbon from lignocellulosic biomass. The process directly converts polysaccharides in lignocellulosic biomass to furan-based precursors for producing fuel-grade hydrocarbons in one step under mild conditions without prior pretreatment and saccharification of the feedstock biomass. Specifically, in the presence of acetone, halide salt, and small amount of mineral acid, the lignocellulosic biomass undergoes hydrolysis of polysaccharides (both cellulose and hemicellulose) to monosaccharides, dehydration of the monosaccharides to furfural (from pentoses) and hydroxymethylfurfural (HMF, from hexoses), and aldol-condensation of HMF and furfural with acetone, and ends up with the formation of furan-derived oxygenates (furfural- and HMF-acetone adducts) with chain length of C5–C15. The resultant furan-based precursors can be subsequently converted to fuel-grade hydrocarbons via hydrogenation. The process integrates the three reactions (Hydrolysis-Dehydration-Aldol condensation) in one step, and is abbreviated as HDA. At the conditions of 120-140°C and 1-2hour, cellulose and hemicellulose in lignocellulosic material (e.g., ground or chipped wood) were completely hydrolyzed into sugars, and more than 80% of the sugars were selectively converted to the furan-acetone adducts. Meanwhile, lignin in the biomass is extensively depolymerized and extracted. Preliminary evaluation indicated the HDA lignin has high purity, low molecular weight with narrow distribution, and good reactivity. The lignin is expected to have great potential in co-products development.

See more of this Session: Chemical and Catalytic Conversions and Processes for Renewable Feedstocks
See more of this Group/Topical: Sustainable Engineering Forum

(471d) Photoanode Area Dependent Efficiency and Recombination In Dye Sensitized Solar Cells

Venkat Kalyan Vendra[1], Jason Absher[1], Samuel Ellis[2], Thad Druffel[2], Delaina A. Amos[1] and Mahendra K. Sunkara[1], (1)Chemical Engineering, University of Louisville, Louisville, KY, (2)Conn Center for Renewable Energy Research, Louisville

Current-voltage characterization was carried out on dye sensitized solar cells (DSSCs) with different photoanode areas ranging from 0.01 cm^2 to 1 cm^2. Surprisingly, we found that the short-circuit current density and the efficiency of the solar cells increased exponentially as the photoaonode area was reduced below 0.1 cm^2. Furthermore, the fill factor remained constant in range of areas investigated and so this behaviour cannot be attributed to the resisitve losses in the solar cell. Understanding the electron transport and recombination in DSSCs with different photoanode areas could hold the key to explain these interesting results.

Several fundamental studies in literature have focused on understanding the electron transport and recombination in nanostructured photoanodes of DSSCs as the film thickness is altered. However, there have not been any reports which investigated how the transport and recombination and efficiency would be affected as the photoanode area of DSSCs varied. Only a few recent studies on organic solar cells have been reported where efficiency of the solar cells was found to dependent on the solar cell area. [1,2] This motivated us to explore how electron transport, recombination are affected with a change in the photoanode area of the DSSC. Interestingly, we found that the transport time constants for these cells of different areas remained the same, but the electron lifetimes increased as the photoanode area was reduced below 0.1 cm^2. In addition, all these studies were performed with photoanodes comprising of different materials such as titania nanoparticles, tin oxide nanowires and tin oxide nanowire and titania nanoparticle composites and an increase in efficiency, short circuit current density and electron lifetime with a decrease in the cell area below 0.1 cm^2 is observed in DSSCs constructed with all these materials. Our results suggest the possibility of different recombination mechanism when the area of the solar cell is reduced below 0.1 cm^2 and also points out the need for a standardized area for characterizing the performance of DSSCs.

[1] Gupta, D., Bag, M., Narayan S. Area dependent efficiency of organic solar cells. Applied Physics Letters 2008, 93, 163301-163304.

[2] Manor, E., Katz, E.A., Tromholt, T., Hirsch, B., Krebs, F.C. Orgin of size effect on the efficiency of organic photovoltaics. Journal of Applied physics 2011, 109, 074508-074517.

See more of this Session: Nanostructured Materials for Dye-Sensitized Solar Cells
See more of this Group/Topical: International Congress on Energy 2011

(235e) Bimetallic Nanoclusters for Glucose Detection

Yixin Liu, Yu Ding and Yu Lei, Department of Chemical, Materials and Biomolecular Engineering, University of Connecticut, Storrs, CT

A nonenzymatic glucose sensor was investigated based on bimetallic nanoclusters modified electrode in this study. The nanoclusters were prepared by combination of electrospinning technique with calcination method. The morphology of the nanoclusters was characterized by scanning electron microscopy and transmission electron microscopy. The structure and composition of nanoclusters was investigated by X-ray diffraction. The electrochemical performance of the as-prepared nanoclusters modified electrode for glucose detection was investigated by cyclic voltammetry and amperometry. Compared to the electrode modified by single metal component, the bimetallic nanoclusters modified electrode showed higher sensitivity towards glucose and better selectivity against ascorbic acid and uric acid. Other analytical characteristics were also reported.

See more of this Session: Micro and Nanofabricated Sensors I
See more of this Group/Topical: Topical 9: Sensors

(286a) Liquid Fuel Production Using Solar-Thermal Energy: Process Development and Technoeconomic Evaluation

Jiyong Kim[1], Terry A. Johnson[2], James E. Miller[3], Ellen B. Stechel[4] and **Christos Maravelias**[1], (1) University of Wisconsin, Madison, WI, (2)Thermal Fluids Science and Engineering, Sandia National Labs, Livermore, CA, (3)Ceramic Processing and Inorganic Materials, Sandia National Labs, Albuquerque, NM, (4)Fuels and Energy Transitions, Sandia National Laboratories, Albuquerque, NM

We have previously described a novel solar-based concept for the production of methanol from carbon dioxide and water. This system applies concentrated solar energy to thermochemically reenergize CO_2 into CO, and then water-gas shift (WGS) to produce syngas (a mixture of CO and H_2) that is fed to a methanol synthesis reactor. Aside from the thermochemical reactor, which is currently under development at Sandia National Laboratories, the approach is based on well-established industrial processes and component designs. Our previous work included an assessment of energy efficiency and economic feasibility of this baseline configuration for an industrial-scale methanol plant. We determined that an overall process energy efficiency (solar-to-fuel) of 7.1% could be achieved and a break-even price of the methanol produced using this approach would be 1.22 USD/kg. Our current work focuses on improvements to the baseline system. We propose several avenues for improvement: 1) alternative pathways to syngas including water splitting and combined CO_2 and H_2O splitting, 2) improved separations, and 3) alternative hydrocarbon products.

For alternative pathways and products, we studied three systems in addition to the baseline system previously reported:
(i) H_2O water splitting, followed by reverse water-gas shift (RWG) reaction, followed by methanol synthesis (MS)
(ii) Simultaneous CO_2 and H_2O splitting (mixed pathway) followed by MS
(iii) Simultaneous CO_2 and H_2O splitting (mixed pathway) followed by Fischer-Tropsch (FT) synthesis

In all designs, we employed amine-based separation systems to remove CO_2 from CO/CO_2 or $H_2/CO/CO_2$ mixtures. However, we also studied the effect of replacing the CO_2 separations system with an analogous CO separations system. We found that CO separations could provide a significant improvement in energy efficiency and utility costs for the CO_2 splitting and mixed pathways.

To compare the improved systems to our baseline, we performed rigorous process simulations to determine equipment sizing and costing, as well as operational expenses (e.g., raw material and utility costs). We then performed detailed profitability calculations. We determined that the mixed pathway (i. e., simultaneous splitting of CO_2 and H_2O) has a number of advantages over the CO_2-splitting or H_2O-splitting pathways. Also, the production of methanol is slightly more attractive than the production of FT diesel.

Using the mixed pathway with MS as our new baseline, we then carried out sensitivity analysis studies to identify the main cost drivers. We found that the primary economic drivers are the high capital investment associated with the solar concentrator/reactor sub-system which accounts for more than 90% of the capital expenditure, and the high utility consumption for CO/CO_2 separation. Economic parameters such as tax rate and interest rate are also important.

Rather than use a cost of CO_2 as had been done previously, we expanded the boundaries of our system to include CO_2 capture at a power plant. We determined that the effect of CO_2 capture leads to small increases in the minimum selling price (i.e., break-even price) of methanol. We also studied how the minimum selling price (MSP) of methanol increases as a function of the distance between the power plant and the solar facility and the transportation technology.

The energy efficiencies of the different systems were also analyzed. Along with lower cost, the mixed pathway with MS demonstrated the highest efficiency. The primary energy efficiency (= chemical energy out (HHV) over total primary energy in) was calculated as 12.7% if process heat and electricity are provided from fossil fuels and 10.7% if solar power is the sole primary energy source. This 10.7% system efficiency is significantly higher than what can currently be achieved with photosynthesis-based processes, and illustrates the potential for solar thermochemical based strategies to overcome the resource limitations that arise for low-efficiency approaches.

Finally, the analysis indicates that a solar-thermochemical pathway to fuels has significant potential, and points towards future research opportunities to increase efficiency and reduce cost. Particularly, it is evident that there is much room for improvement in the development of a less expensive and more efficient solar concentrator/reactor sub-system, an opportunity that will benefit from the increasing deployment of concentrated solar power. In addition, significant advances are achievable through improved separations, different end products, and different degrees of process integration and distribution. Our investigation establishes a methodology for comparison and assessment of impact on both the economics and the efficiency for advanced system designs.

"This work was supported by the Laboratory Directed Research and Development program at Sandia National Laboratories, in the form of a Grand Challenge project entitled Reimagining Liquid Transportation Fuels: Sunshine to Petrol. Sandia is a multiprogram laboratory operated by Sandia Corporation, a Lockheed Martin Company, for the United States Department of Energy's National Nuclear Security Administration under Contract DE-AC04-94AL85000. "

(355d) Non-Biological Inhibition Based Sensing (NIBS) for Detection of Trihalomethanes (THMs) In Drinking Water

Isaac K. Afreh and Chelsea N. Monty, Chemical and Biomolecular Engineering, University of Akron, Akron, OH

Abstract on Non-biological Inhibition Based Sensing (NIBS) for Detection of Trihalomethanes (THMs) in Drinking Water

The use of chlorine for disinfection is a very popular and effective method for drinking water treatment. These chlorine disinfectants are able to destroy waterborne pathogens and also protect drinking water from contamination during delivery to consumers. However, chlorine reacts with natural organic matter (NOM) and anthropogenic chemicals present in the water to form disinfection by-products (DBPs), such as trihalomethanes (THMs), which can be carcinogenic, genotoxic, cytotoxic, and hepatotoxic at moderate concentrations. Unfortunately, there are currently no "real-time" techniques available to monitor for low levels of THMs in finished drinking water. Many drinking water treatment facilities must send samples to off-site laboratories in order to determine DBP concentrations. This can create a delay in detection of up to 48 hours and result in the distribution of potentially toxic drinking water to consumers. Therefore developing a detection method that will rapidly sense low concentrations of regulated THMs in aqueous samples would help mitigate potential health concerns and increase the quality of water being delivered to consumers' homes. Our solution is to develop a sensor that will provide a "real-time" detection of toxic THMs in drinking water. The chemistry for this sensor is based on a new technique in chemical amplification called non-biological inhibition based sensing (NIBS). In NIBS, instead of the analyte (THM) acting as a catalyst, the analyte rather acts as an inhibitor for the given reaction. The amount of the THM present in the reaction is then determined from the kinetics of the inhibited NIBS reaction. The catalyst solution for the reaction is prepared via Fujiwara reaction, where the THM (inhibitor) binds to the pyridine (catalyst). In this presentation, I will show how conditions (such as temperature, time, phase and concentration of THMs) under which the Fujiwara reactions are run affect inhibition by the THMs. Also, I will discuss the effect of temperature on the kinetics of both uninhibited and inhibited NIBS reactions.

See more of this Session: Catalytic, Environmental, and Industrial Gas Sensors
See more of this Group/Topical: Topical 9: Sensors

(288a) Managing Your Career

Wendy Young, GE Water & Process Technologies, El Lago, TX

Powerpoint to be added later

See more of this Session: Developing Leadership Skills
See more of this Group/Topical: Management Division

(11e) Ultra-Low Fouling Zwitterionic Carboxybetaine Antibody Microarrays for Cancer Diagnostics Directly From Human Blood

Norman D. Brault, Qiuming Yu and Shaoyi Jiang, Chemical Engineering, University of Washington, Seattle, WA

Cancer biomarkers play a vital role in early disease diagnostics, monitoring the therapeutic response, and improving patient stratification for treatment. The extreme complexity and heterogeneity of cancer both between patients and between cells within a patient has lead to a continued push for discovering more of these biological indicators. This has resulted in the use of high-throughput antibody-based biosensor platforms which can detect multiple disease biomarkers simultaneously in order to enhance the sensitivity and accuracy of diagnostic assays. However, the presence of non-specific protein adsorption on biosensors has been the major culprit preventing the highly sensitive detection of these target analytes (at or below nanogram/milliliter quantities) from the complex environment of human blood serum or plasma. In this work, antibody arrays were created using a simple yet robust zwitterionic poly(carboxybetaine acrylamide) (pCBAA) surface platform. The functionalization capabilities combined with the ultra-low protein fouling properties both before and after effective immobilization of biorecognition elements made pCBAA highly desirable. Using standard pin-spotting techniques on the pCBAA films, arrays containing antibodies to multiple cancer biomarkers were analyzed using a highly sensitive self-referencing surface plasmon resonance (SPR) imaging sensor with polarization contrast. Successful antibody immobilization and subsequent reduced biofouling properties of the functionalized surfaces resulted in a dynamic detection range with high sensitivity and reproducibility for the simultaneous detection of multiple cancer biomarkers directly from undiluted human blood plasma. The lack of cross reactivity and high selectivity enabled these coatings to be used with actual cancer patient samples illustrating the great potential of this new biosensing platform.

See more of this Session: Biosensor Devices I
See more of this Group/Topical: Topical 9: Sensors

Determination of kinetic parameters of self-ignition wood pellets during storage

Wendi Guo[1], Xiaotao Bi[1], C. Jim Lim[1], Shahab Sokhansanj[1,2]

1. Chemical & Biological Engineering, University of British Columbia, Vancouver, BC V6T 1Z3
2. Environmental Sciences Division, Oak Ridge National laboratory, Oak Ridge TN 37831

1. Introduction

Accidental fires in wood pellets silos happen frequently (Channel 3000 2007; Aines 2008; Bartel 2008). It appears that the oxidative biological reactions are initially responsible for simultaneous gas release and temperature increase. (Walker and Harrison 1977; Lehtikangas 2000; Wihersaari 2005b, 2005a). The off-gassing and self-heating phenomena of wood pellet during storage have been investigated by several research institutions as well as in our Biomass & Bioenergy laboratory at UBC (Svedberg *et al.* 2004; Wadso 2007; Kuang *et al.* 2008; Svedberg *et al.* 2008; Kuang *et al.* 2009a; Kuang *et al.* 2009b; Guo *et al.* 2011). The results to date show that the rate of self-heating decreases with the age of pellets. The reaction rate is also highly dependent on storage temperature. A maximum heat release rate of 0.3 W/(kg pellet) has been recorded when a mass of pellets was stored in a sealed container maintained at 50 °C.

It is postulated that chemical oxidation dominates the self heating process when the storage temperature increases to 80 °C. This process leads to a higher temperature accelerating self heating when a self-ignition may occur (Meijer and Gast 2004). The self ignition of wood and other biomass was widely investigated experimentally under heat flux. Models were developed to predict the thermal ignition for those materials (Bilbao *et al.* 2001; Bilbao 2002; Pauner and Bygbjerg 2007)

The kinetic of wood pellets self heating at higher temperatures and the self-ignition phenomena haven't been fully understood. Experiments were carried out in this work at temperature range of 100 °C to 200 °C to study wood pellets chemical oxidation process. A self-heating model was developed to simulate the self-heating process using the results obtained from the experiment. A numerical method was developed to solve the energy balance and obtain the kinetic parameters for self-heating process.

2. Methodology

Experiments were carried out using a lab size oven (Thermal scientific, 1.7 cu. ft). Cylindrical containers were used as the self heating reactor and were put in the oven (constant temperature) with wood pellets loaded. Assuming the self heating process is a first order reaction and the self heating heat can be described as reaction heat (enthalpy of reaction, $\Delta_r h$) multiplying the reaction rate constant [$k=A\exp(-E/RT)$].

The energy balance of the self-heating process inside a cylindrical container is expressed by

$$\rho C_P \frac{\partial T}{\partial t} = \lambda(\frac{1}{r}\frac{\partial}{\partial r}(r\frac{\partial T}{\partial r}) + \frac{\partial^2 T}{\partial z^2}) + \rho \cdot \Delta_r h \cdot A \exp(-\frac{E}{RT}), \quad (1)$$

where, C_p is the specific heat capacity of the pellets, kJ/kg; λ is the thermal conductivity of the pellets , W/(m K); ρ is the bulk density of the pellets, kg/m^3; T is the temperature of the pellets, K; r is the radius of the container, m; $\Delta_r h$ is the reaction enthalpy of the self-heating reaction, kJ/mol; A is a exponential factor; E is the activation energy of the self-heating reaction, J/mol; R is gas constant, 8.314 J/(molK); t is time of the self heating reaction, s.

A curve-fitting method was developed to extract the kinetic parameters (including $\Delta_r h$, A and E) using the experimental data (temperature vs. time). Eq.(1) was solved by MatLab using the thermal properties obtained from the previous studies along with the boundary conditions shown in Figure 1. The best fitted kinetic parameters were determined by minimizing the residuals between the numerical solution and the experimental temperature profiles.

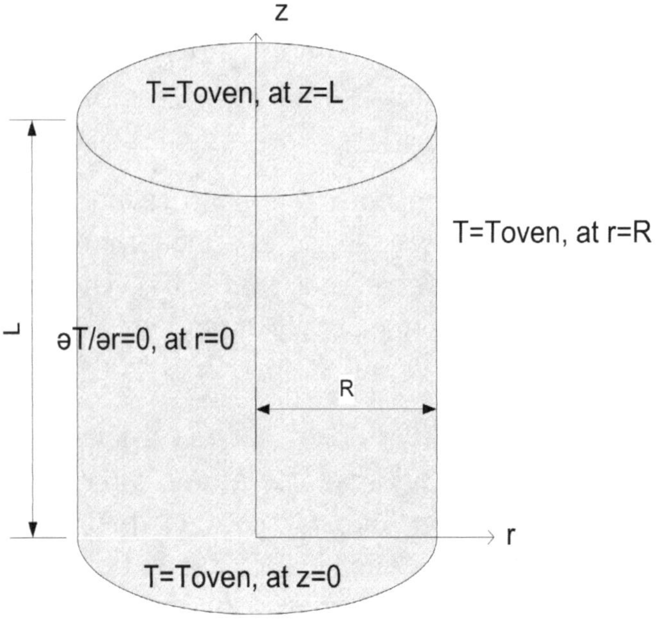

Figure 1 Boundary conditions and assumptions

Assumptions:

Bulk pellets can be treated as homogeneous materials with effective thermal properties (Cp, λ, ρ).

Initial condition t=0, $T=T_0$ (initial temperature)

No air movement inside the container

Fast heat conduction at container wall ($T_{wall}=T_{oven}$)

3. Experiment setup and materials

3.1 Sample

Softwood pellets produced in British Columbia (BC) were used in this research. Pellets with designation A, B and C in this research were manufactured from 3 different wood pellet plants in BC. They were made of whitewood mostly pine; both pellet A and pellet C were manufactured by Princeton Co-Generation Corp. (Princeton, BC, CanadaV0X 1W0). Pellet B was manufactured by Premium Pellet Ltd. (Vanderhoof, British Columbia Canada, V0J 3A0). Pellets were on average 6.2 mm in diameter and 10 to 25 mm in length. All pellets were made using whitewood with no bark. The pellets in this region are made from sawdust and shavings, a by-products of milling spruce, pine and fir for lumber.

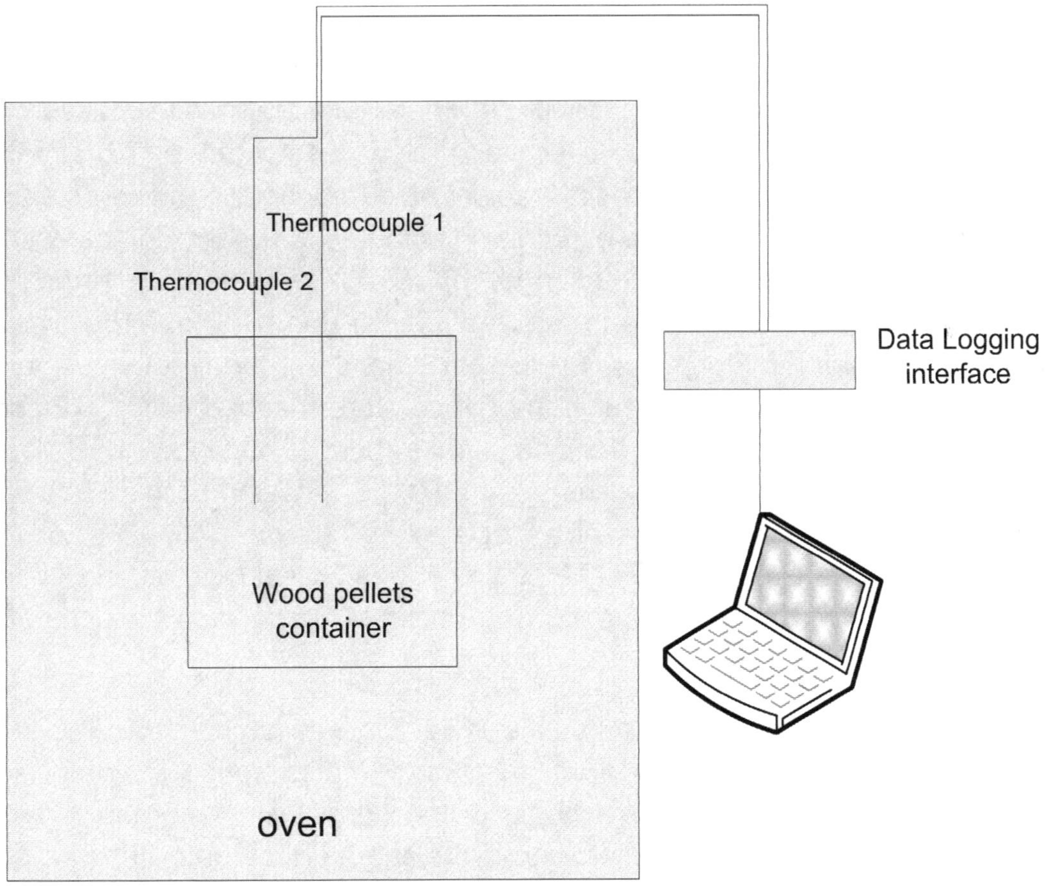

Figure 2 Experimental setup

3.2 Experimental setup

Figure 2 shows the experimental setup. Three sizes of containers made from steel were used in this experiment, small (H 106mm, D 120mm); Medium (H 166mm, D 190mm); Large (H 226mm, D 240mm). Top open short cylinders were used as reactor and two thermocouples were placed inside the container, one right in the middle (T1) and another one half way to the center (T2) (radial). The ends of two thermocouples were kept at the same axial level (same distances to the top and bottom). After placing the thermocouples, the containers were filled up with wood pellets (weight was recorded) and put inside of an oven (maintained at a constant temperature (between 120°C to 200°C); temperatures were recorded using a data acquisition system from the start till after the crossing point.

4 Results and Discussion

4.1 Oven temperature and self-heating

Figure 3 plots the experimental results of two different size containers at different oven temperature. The experimental results show that there is no self heating detected below an oven temperature of 100 °C at all lab size self heating containers which used in this study. Figure 3.a shows the temperature profiles of the large container (L)at 103 °C for 4 days and no self heating was detected. Self-heating was detected when the oven temperature was set at 140 °C for large containers (L), as shown in Figure3b where temperatures increased with time. When the center temperature (T1) crossed over the off-center temperature (T2) at 135°C (the crossing point), both temperature increased rapidly and reached a higher steady state temperature. Here, we consider the CP point (crossing point) as a signal of the detectable self-heating. When the oven temperature increased to 150 °C, the temperature T1 and T2 of the large container (L) continued increasing and reached 158 °C (the peak of T1) a steady state (decrease slightly), as shown in Figure 3.c. On the other hand, for the medium container, the crossing point temperature didn't appear until the temperature reached 147 °C (Figure 3.c.). For the large container, there is finally an uncontrollable thermal run-away which leads to a self-ignition when the oven temperature increased to 170 °C (shown in Figure3.d), while the temperature of medium container remains at a steady state (around 175 °C) . For the large container, temperatures kept running away after the oven heat was turned off and led to a self-ignition at 200 °C. Figure 3.f shows the burning wood pellets. The ignition happens at a point near the center.

It can be concluded that the heat release of the oxidization at temperature 100-200 °C is significant enough to be detected using a lab technique and to cause a self-ignition. The crossing point (detectable point of self-heating) and the thermal run-away temperatures are highly dependent on the size of the containers and the temperature of the oven.

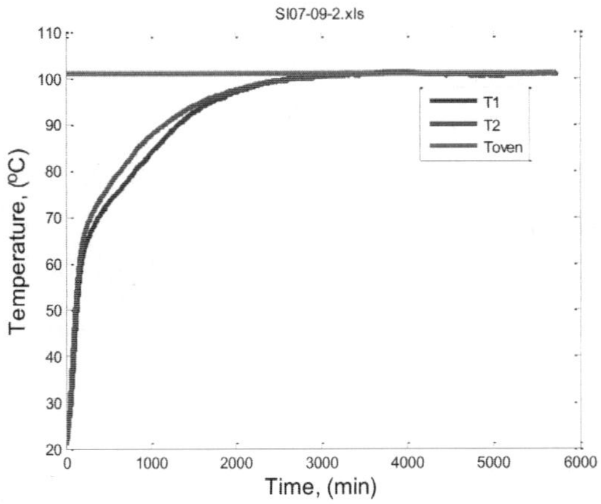

a.Oven temperature 103 °C, no crossing point, no thermal run away.

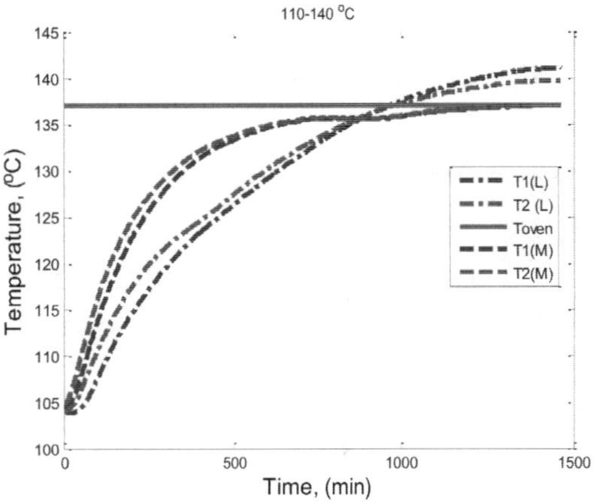

b. Oven temperature 140 °C crossing point after 10 minutes (Large and Medium container)

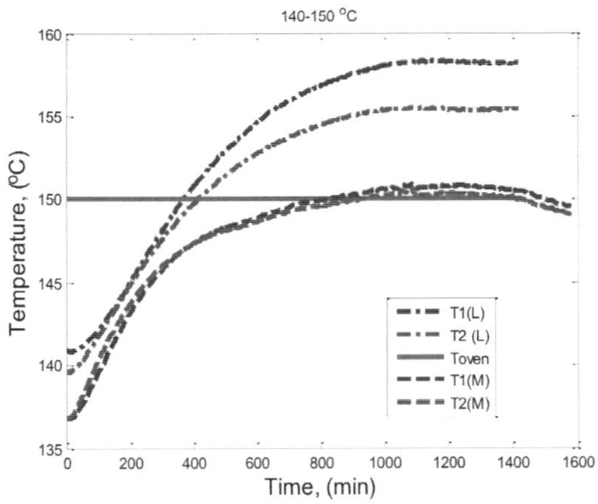

c. Oven temperature 150 °C (L&M)

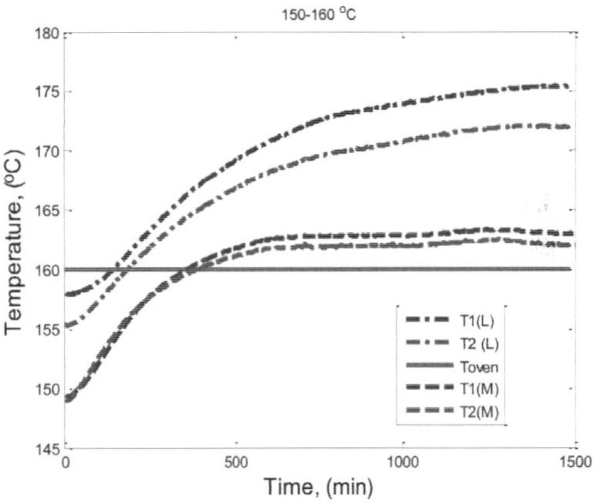

d. Oven temperature 160 °C (L&M)

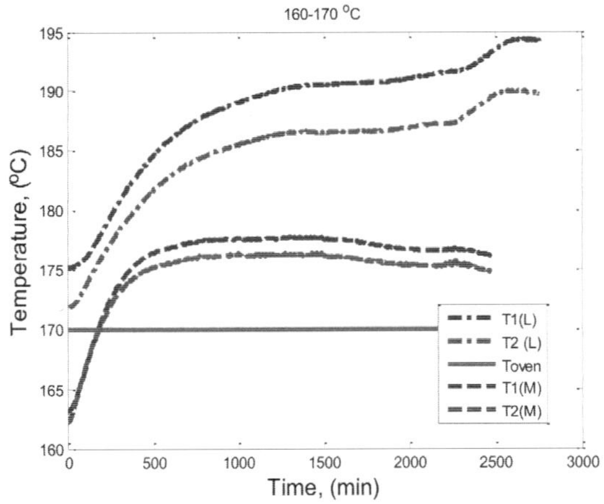

e. Oven temperature 170 °C and self-ignition (L&M).

f. Burning wood pellets.

Figure 3 Self-heating and self-ignition -Temperature profiles at different oven temperatures.

4.2 Kinetic parameters from curve fitting

A curve-fitting method was developed by solving the energy balance equation Eq (1) based on the boundary conditions shown in Fig. 1. The kinetic parameters were obtained by searching the best fit using the curve-fitting method and both data from large containers and medium containers. Figure 4 shows the typical best fitted curves in comparison with the experimental data.

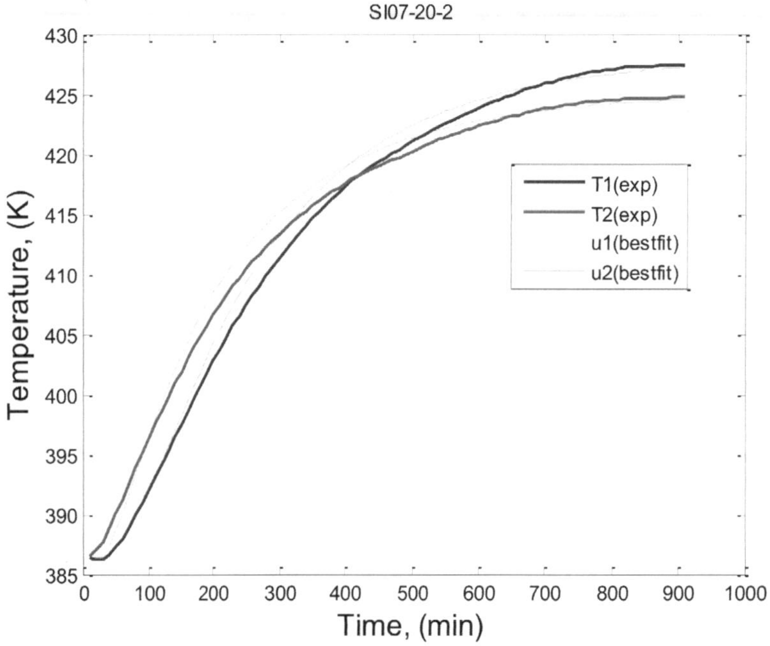

Figure 4 Comparison of the experimental data and best fitted curves of the kinetic parameters. (Experimental results of 'SI07-20-2' comparing to the numerical results of parameter E=62.5 kJ/mol, $\rho\Delta_r hA=5.7*10^7$ J.K/(m^3.s)

Further analysis is in progress.

5. Summary

Experiments were carried out to investigate the self-heating and self-ignition process of wood pellets at temperatures ranging from 100°C to 200°C. Three sizes of containers were used. The results show that the self-heating speed and thermal run-away temperature are highly dependent on the oven temperature and the size of the reactors. A curve-fitting method is developed to extract the kinetic parameters.

Aines, D. 2008. Fire reported for second time in six weeks at wood stove pellet plant. Retrieved 08 July 2010, from http://www.herald-mail.com/?cmd=displaystory&story_id=201753&format=html.

Bartel, R. 2008. Fire at westwood fibre products Retrieved 08 July 2010, from http://www.silk.fm/node/763122.

Bilbao, 2002. A model for the prediction of the thermal degradation and ignition of wood under constant and variable heat flux. Journal of Analytical and Applied Pyrolysis **62**(1): 63.

Bilbao, R., Mastral, J.F., Aldea, M.E., Ceamanos, J., Betr, M., Lana, J.A., 2001. Experimental and theoretical study of the ignition and smoldering of wood including convective effects. Combustion and Flame **126**(1-2): 1363-1372.

Channel 3000. 2007. 2 injured in explosion, fire at wood pellet warehouse. Retrieved 08 July 2010, from http://www.channel3000.com/news/10947450/detail.html.

Guo, W., Trischukb, K., Bi, X., Lim, C.J., Sokhansanj, S., 2011. Measurements of wood pellet self-heating kinetic parameters using isothermal calorimetry. Submitted to Biomass & Bioenergy. **Under Review**.

Kuang, X., Shankar, T.J., Bi, X.T., Lim, C.J., Sokhansanj, S., Melin, S., 2009a. Rate and peak concentrations of off-gas emissions in stored wood pellets—sensitivities to temperature, relative humidity, and headspace volume. Annals of Occupational Hygiene **53**(8): 789-796.

Kuang, X., Shankar, T.J., Bi, X.T., Sokhansanj, S., Jim Lim, C., Melin, S., 2008. Characterization and kinetics study of off-gas emissions from stored wood pellets. Annals of Occupational Hygiene **52**(8): 675-683.

Kuang, X., Shankar, T.J., Sokhansanj, S., Lim, C.J., Bi, X.T., Melin, S., 2009b. Effects of headspace and oxygen level on off-gas emissions from wood pellets in storage. Annals of Occupational Hygiene **53**(8): 807-813.

Lehtikangas, P., 2000. Storage effects on pelletised sawdust, logging residues and bark. Biomass Bioenerg. **19**(5): 287-293.

Meijer, R., Gast, C.H., 2004. Spontaneous combustion of biomass: Experimental study into guidelines to avoid and control this phenomena. KEMA Power Generation & Sustainables.

Pauner, M., Bygbjerg, H., 2007. Spontaneous ignition in storage and production lines: Investigation on wood pellets and protein powders. Fire and Materials **31**(8): 477-494.

Svedberg, U.R.A., Hogberg, H.E., Hogberg, J., Galle, B., 2004. Emission of hexanal and carbon monoxide from storage of wood pellets, a potential occupational and domestic health hazard. Annals of Occupational Hygiene **48**(4): 339-349.

Svedberg, U.R.A., Samuelsson, J., Melin, S., 2008. Hazardous off-gassing of carbon monoxide and oxygen depletion during ocean transportation of wood pellets. Annals of Occupational Hygiene **52**(4): 259-266.

Wadso, L., 2007. Measuring chemical heat production rates of biofuels by isothermal calorimetry for hazardous evaluation modelling. Fire and Materials **31**(4): 241-255.

Walker, I., Harrison, W., 1977. Self-heating of wet wood. 1. Exothermic oxidation of wet sawdust. New Zealand Journal of Science **20**: 191-200.

Wihersaari, M., 2005a. Evaluation of greenhouse gas emission risks from storage of wood residue. Biomass Bioenerg. **28**(5): 444-453.

Wihersaari, M., 2005b. Greenhouse gas emissions from final harvest fuel chip production in finland. Biomass Bioenerg. **28**(5): 435-443.

(76e) Quasi-3D Plasmonic Nanostructures for Strain Specific Identification of Marine Bacteria Vibrio Parahaemolyticus Using SERS

Jiajie Xu[1], Matthew Idso[1], Qiuming Yu[1], Jeff Turner[2] and Mark S. Strom[2], (1)Chemical Engineering, University of Washington, Seattle, WA, (2)Northwest Fisheries Science Center, National Oceanic and Atmospheric Administration, Seattle, WA

Surface-enhanced Raman spectroscopy (SERS) enabled by the advance of nanotechnology has immerged as a powerful analytical and sensing tool for detecting a verity of analytes from toxins to entire microorganisms. The extreme strong local electric fields, raising from the local surface plasmon resonance (LSPR), in the vicinity of noble metallic nanostructures can significantly increase the detection sensitivity by 10^6 -10^{14}-fold, which enables the detection of single molecules.[1,2] Since the strength of the electromagnetic field decays exponentially from the metal surface, SERS is limited to the near-surface of diffusely scattering objects. As a consequence, the SERS technique is well suited for probing cell membrane architectures and constituents that directly contact the surface of metal nanostructures. Several studies have demonstrated the capability of SERS for distinguishing various bacteria at the species-level by either mixing bacteria with colloidal nanoparticle suspensions or dropping bacteria on SERS-active substrates.[3,4] However, for many important applications such as food safety, environment monitoring, and homeland security, the strain-level identification is desired due to the large variation of pathogenicity among strains.

Here, we present the ***in situ*** identification of different strains of marine bacterium *Vibrio Parahaemolyticus* using SERS based on the unique quasi-3D plasmonic nanostructure arrays developed in our group. The quasi-3D nanostructure arrays are composite of physically separated gold thin film with nanoholes on top and gold nanodisks at the bottom of the wells. Unlike previous SERS substrates of random gold nanoclusters, these SERS substrates are fabricated via electron beam lithography (EBL) which allows us to optimize nanostructures for SERS sensitivity of large organisms, and to precisely control the dimension that enables quantitative and reproducible analysis. We conducted finite-different time-domain (FDTD) electromagnetic calculations to determine the optimal nanostructures for detecting microorganisms. The optimal structure, which is 400 nm in diameter, 100 nm in spacing and 400 nm in depth, was fabricated via EBL followed by evaporation of 50 nm gold. Seven *Vibrio Parahaemolyticus* strains obtained from environment including oyster, plankton and water as well as from clinical patients stool were selected to set up an *in situ* bacteria "fingerprint" database.

Vibrio Paraheamolyticus anabiosis cultures were grown 12-16 hours in tryptic soy broth (TSB) at 37 °C with shaking to achieve exponential-phase growth. Bacteria solution was washed for 3 times with 1 g/L of 0.5% NaCl and then diluted to 10^8 CFU/ml. The SERS-active substrate with four 50 μm x 50 μm quasi-3D nanostructure arrays was immersed in the bacteria solution in a custom-built Teflon cell covered with a piece of cover glass, allowing the collection of *in situ* SERS spectra of bacteria. Total 32 spectra were collected on each substrate and two substrates were used for each strain. Results showed highly reproducibility due to the well controlled dimension of the nanostructure arrays.

All the raw spectra were baseline-corrected using cubic spline interpolation. The intensity and wavenumber of 25 main peaks of baseline-corrected spectra were used for principle component analysis (PCA) [5,6] to generate the principle components (PCs) translation matrix. The first three PCs (PC1, PC2, and PC3), which account for greater than 90% of the variance in the analysis, were used to characterize these spectra and to differentiate the *Vibrio Parahaemolyticus* strains. All seven strains can be clearly differentiated in the 3D-PCA plot. Furthermore, we conducted the blind test by randomly selecting three strains from the seven tested strains. Following the same method, the PCA values of three blind samples overlap with the clusters of three known strains, and thereby, the strains of the blind samples were determined. The successful determination of the strains of blind samples indicate that quasi-3D plasmonic nanostructure arrays can be used as SERS-active substrates for strain specific identification of marine pathogens.

(1) Kneipp, K.; Wang, Y.; Kneipp, H.; Perelman, L. T.; Itzkan, I.; Dasari, R. R.; Feld, M. S. *Phys. Rev. Lett.* **1997**, *78*, 1667.

(2) Nie, S.; Emory, S. R. *Science* **1997**, *275*, 1102.

(3) Patel, I.; Premasiri, W.; Moir, D.; Ziegler, L. *J. Raman Spectrosc.* **2008**, *39*, 1660.

(4) Jarvis, R.; Brooker, A.; Goodacre, R. *Faraday Discuss.* **2006**, *132*, 281.

(5) Xu, J. J.; Zhang, L.; Gong, H.; Homola, J.; Yu, Q. M. *Small* **2011**, *7*, 371.

(6) Jolliffe, I. *Principal component analysis*; Springer verlag, 2002.

See more of this Session: Biosensor Devices II
See more of this Group/Topical: Topical 9: Sensors

MODELLING SIMULTANEOUS SACCHARIFICATION AND FERMENTATION OF MICROCRYSTALLINE CELLULOSE USING COMPUTATIONAL FLUID DYNAMICS

Josebus Maree van Zyl, Stellenbosch University, Stellenbosch, RSA
Thomas Michael Harms, Stellenbosch University, Stellenbosch, RSA
Eugéne van Rensburg, Stellenbosch University, Stellenbosch, RSA
Willem Heber van Zyl, Stellenbosch University, Stellenbosch, RSA
Lee Rybeck Lynd, Dartmouth College, Hanover, NH

Presentation ID#: 233146

Biography

J.M. van Zyl
email: jmvz@sun.ac.za
Tel: +27 (0)21 808 4272

Extended Abstract

Cellulose is a polysaccharide and one of the most abundant organic compounds found on the Earth surface [1]. It holds the potential to reduce human reliance on fossil derived fuels by producing ethanol through the hydrolysis and fermentation of cellulose. Ethanol is a highly combustible liquid fuel which has been shown to be a viable alternative to petrol. Producing ethanol from cellulose however has proved most challenging, particularly overcoming the recalcitrance of cellulose to enzymatic hydrolysis.

Cellulose to ethanol conversion has successfully been demonstrated at laboratory scale for a variety of feedstock and production methods and in a few cases at demonstration scale. Commercialization of cellulosic ethanol is however limited by the lack of large capital funds required to construct such facilities. A compounding effect is that investors still consider the technology as immature and high risk and as such are hesitant to commit. Obtaining the necessary capital funding requires de-risking the technology as well as policy frameworks and incentives from government.

Engineers have recently begun the development of numerical models for implementation with computational fluid dynamic (CFD) software that will allow investigation of secondary effects of cellulosic ethanol production scale-up. The advantage of CFD is the modelling of the flow patterns and conditions that could exist in a commercial scale fermenter. This includes shear-rates, power requirements, degree of particle suspension and through user-coded modules the reaction-rates of biological or chemical processes. This methodology could allow a significant reduction in the risk associated for investors at a reasonably low financial expense.

Presented in this paper are results of the reaction kinetic modelling (Figure 1) of microcrystalline cellulose to ethanol conversion using cellulases from the fungus *Trichoderma reesei* and an industrial grade yeast *Saccharomyces cerevisiae*, commonly used in the food industry [2]. Observations from the results indicated that the number of available bonding sites on the particle surface appeared to control the rate at which hydrolysis occurred. Enzymes appear to catalyse a reaction at a constant rate thus more sites available to attach should result in higher reaction rates. However due to the fibril structure of microcrystalline cellulose a large number of bonding sites are available resulting in a large number of exoglucanase and endoglucanase enzymes initially attaching to sites on the cellulose chains. Once the initial available sites diminish, a large percentage of the exoglucanase enzymes detach from the particle surface and enters the fermentation broth, effectively reducing the rate of hydrolysis. It thus suggests that if the work rate of the endoglucanase enzyme could be significantly increased, high hydrolysis rates could be achieved as more reduction ends become available for the exoglucanase enzymes.

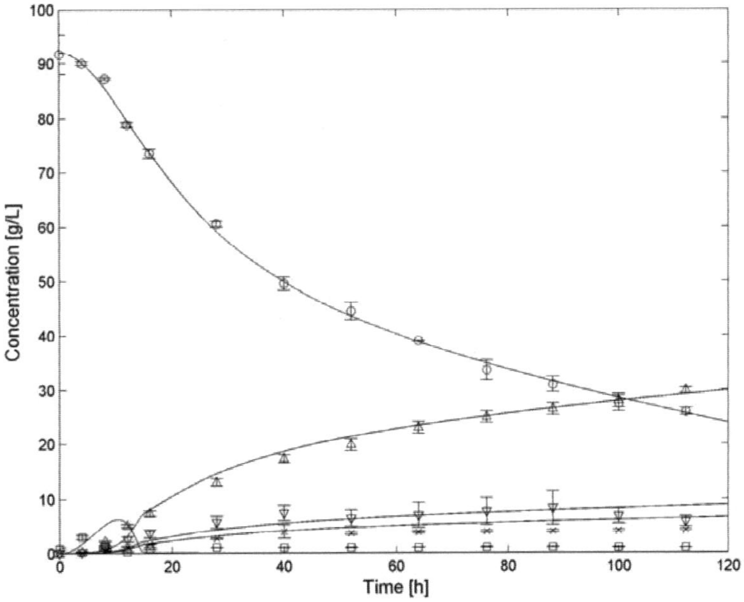

Figure 1: Kinetics model for hydrolysis and fermentation of Avicel PH-101
SSF of Avicel (○) forming glucose (□) fermented to ethanol (Δ) and glycerol (x). Additional yeast cells grown (∇). Simulated results are presented with solid lines.

Microcrystalline cellulose is an insoluble compound which needs to be modelled as particles. It is further required to determine the effects of each substance in the fermentation broth on the viscosity of the system. Results indicated microcrystalline particles occur in various shapes and sizes and exhibit significantly higher drag resistance when the average particle sized is in the order of 50 μm and smaller. This effect is possibly due to the combination of particle topology, structured water [3] and near surface effects [4]. Furthermore the presence of particles has the greatest influence on the apparent dynamic viscosity of the fermentation broth (Figure 2). A numerical model was developed for use in CFD codes which could predict the effects of the particles along with ethanol and glycerol on the apparent dynamic viscosity of the fermentation broth with an average error of 11.1 %.

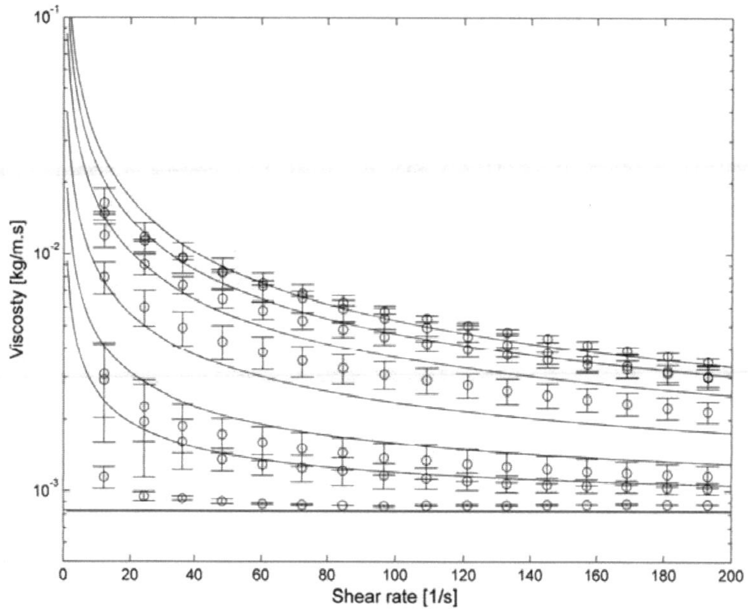

Figure 2: Dynamic viscosity influence of Avicel PH-101 particles

Experimental results (○) with error-bars indicating the standard deviation compared to the numerical predictions (–)

The above mentioned models were incorporated into the commercial CFD package STAR-CCM+ [5] to evaluate the dominant flow patterns and particle distribution within a typical batch mixing vessel. The kinetic models were used to calculate the efficiency of the mixing vessel in-terms of ethanol production rate based on the substrate and reactants concentrations. Results indicated that as cellulose conversion occurs the dynamic viscosity of the fermentation medium increases while the particle sizes decrease. This leads to improved particle suspension and homogeneity which in-turn approaches the optimal conditions for biological hydrolysis and thus ethanol production.

References

1. Lynd LR, Weimer PJ, van Zyl WH, and Pretorius IS, 2002, Microbial cellulose utilization: Fundamentals and Biotechnology, Microbiology and Molecular Biology Reviews, 66 (3): 506-577.
2. van Zyl JM, van Rensburg E, van Zyl WH, Lynd LR, and Harms TM, 2010, A kinetic model for simultaneous saccharification and fermentation of avicel with saccharomyces cerevisiae, Biotechnology and Bioengineering. 108:924-933.
3. J.F. Matthews, C.E. Skopec, P.E. Mason, P. Zuccato, R.W. Torget, J. Sugiyama, M.E. Himmel, J.W. Brady, Computer simulation studies of microcrystalline cellulose Iβ, Carbohyd. Res. 341 (2006) 138-152.
4. J. Zheng, G.H. Pollack, Long-range forces extending from polymer-gel surfaces. Phys. Rev. E, 68 (2003) 031408.
5. STAR-CCM+, version 6.02.007, CD-Adapco, London, UK

(286d) Design and Optimization of Solar Thermal Systems

Amin Ghobeity, Corey J Noone, Enrique Lizarraga-Garcia and **Alexander Mitsos**, Mechanical Engineering, Massachusetts Institute of Technology, Cambridge, MA

Solar-thermal systems based on Concentrated Solar Power (CSP) are considered a promising method to generate electricity from solar energy. CSP systems have been considered for several decades and in recent years generate renewed interest in research, development and commercialization. Several difficulties associated with CSP systems have to be overcome before a successful wide-scale commercial adoption.
The focus of this talk is the design and optimization of CSP with a central receiver, in particular integrated with thermal energy storage. Such technologies are still at a research/development stage.

Central receiver systems are typically employed in flat land using dozens to thousands of heliostats for a single receiver. In the first part of the talk we describe a recent proposal to use hilly terrain [1,3]. Hilly terrain is often significantly cheaper than flat lands which is suitable for many uses. A tool for site selection is presented for both beam-up and beam-down configurations [1] and used to demonstrate the abundance of suitable hills. Additionally a computationally efficient tool for heliostat placement in both planar and hilly terrains is described and used to identify novel heliostat patterns that are significantly better than any of the existing ones [2].

An interdisciplinary team at MIT proposed a new concept, termed CSPonD, that collocates the receiver and storage system [3]. CSPonD has the potential to overcome some of the operational challenges of power towers and reduce capital cost of the receiver/storage system without compromising the efficiency. On the other hand, identifying an optimal design and operation is very challenging. The second part of the talk describes a methodology based on dynamic optimization [4,5,6]. Both single-purpose (electricity only) and dual-purpose plants (electricity and seawater desalination) are discussed. The effect of feed-in tariff (FIT) on optimal design and operation is demonstrated and the influence of seasonal variations on operability discussed. Finally, different energy storage systems (both conventional and novel) are compared in terms of thermodynamic efficiency and cost.

[1] C. J. Noone, A. Ghobeity, A. H. Slocum, G. Tzamtzis and A. Mitsos. Site Selection for Hillside Central Receiver Solar Thermal Plants, Solar Energy, 85 (5):839-848 2011.

[2] C. J. Noone and A. Mitsos, Heliostat Layout Optimization via Automatic Differentiation: Adjoints and Convex Relaxations. SIAM Computational Science and Engineering Conference, Reno, NV, 2011.

[3] A. H. Slocum, D. S. Codd, J. Buongiorno, C. Forsberg, T. McKrell, J.-C. Nave, C. N. Papanicolas, A. Ghobeity, C. J. Noone, S. Passerini, F. Rojas and A. Mitsos. Concentrated Solar Power on Demand,

In Press: Solar Energy, April 11, 2011.

[4] A. Ghobeity and A. Mitsos, Optimal Use of Solar Thermal Energy for Combined Power Generation and Water Desalination (Plenary Talk, Paper #174), Nicosia, Cyprus. Conference on the Promotion of Distributed Renewable Energy Sources in the Mediterranean Region (DISTRES), 2009.

[5] A. Ghobeity and A. Mitsos. Optimal Operation of a Concentrated Solar Thermal Cogeneration Plant. 21st European Symposium on Computer Aided Process Engineering ESCAPE 21. Chalkidiki, Greece. 2011

[6] A. Ghobeity, E. Lizarraga-Garcia and A. Mitsos. Optimal Design and Operation of a Volumetric Solar-Thermal Energy Receiver and Storage. ECOS 2011, The 24th International Conference on Efficiency, Cost, Optimization, Simulation and Environmental Impact of Energy Systems, Novi Sad, Serbia, 2011.

See more of this Session: Concentrated Solar for Power Generation and Chemical Processing I
See more of this Group/Topical: International Congress on Energy 2011

(144a) Asynchronous Magnetic Bead Rotation In Microfluidic Droplets for Monitoring Cell Growth

Irene Sinn[1], Paivo Kinnunen[2], Theodore Albertson[3], Brandon H. McNaughton[1], Mark A. Burns[4] and Raoul Kopelman[5], (1)Biomedical Engineering, University of Michigan, Ann Arbor, MI, (2)Applied Physics, University of Michigan, Ann Arbor, MI, (3)Physics, University of Michigan, Ann Arbor, MI, (4)Chemical Engineering, University of Michigan, Ann Arbor, MI, (5)Chemistry, University of Michigan, Ann Arbor, MI

The rapid identification and characterization of bacterial pathogens is an important goal for point-of-care diagnostics. Droplet-based lab-on-a-chip systems decrease analysis time and allow for higher levels of throughput because each droplet acts as an isolated reaction chamber. We demonstrate an asynchronous magnetic bead rotation (AMBR) droplet system for single cell and colony growth studies. By monitoring a magnetic bead's asynchronous frequency response to a rotating magnetic field, the continuous growth and division of individual bacterium and the growth of a small bacterial colony can be observed. The sensitivity of the combined AMBR droplet microfluidics system reduces minimum inhibitory concentration (MIC) measurement time from days to minutes. This technology provides a single cell, as well as a colony analysis platform, that can be used for rapid, high-throughput characterization of individual bacteria or bacterial colonies.

See more of this Session: Biosensor Devices III
See more of this Group/Topical: Topical 9: Sensors

(258f) Flow and Transport Properties In High-Solids Biomass Slurries

James J. Lischeske and Jonathan J. Stickel, National Bioenergy Center, National Renewable Energy Laboratory, Golden, CO

Economic biochemical conversion of cellulose to sugars necessitates minimal addition of water to biomass slurries during conversion, resulting in high-solids systems. The resulting slurries are effectively wet-granular materials. These materials presents difficulties for reactor engineering, as their flow and transport properties, along with the mechanical properties that govern them, are poorly understood. This work examines the nature of high-solids flow, along with the heat and mass transfer properties of biomass and related model materials. Methods were developed to measure the volume fraction of solid, liquid and gaseous phases of high-solids slurries. The effect of phase composition on flow was systematically examined in a rotating vessel. Qualitative observations, as well as accessible quantitative information about size distribution of aggregates, were made. Further, a model for measuring bed diffusion in biomass was developed, and experiments were performed to measure the bed diffusion coefficient in model materials and biomass. Finally, heat transfer in a packed bed of biomass was examined, and conductivity coefficients were calculated and related to phase composition.

See more of this Session: Reaction Kinetics and Transport Fundamentals for Biomass Conversion
See more of this Group/Topical: Sustainable Engineering Forum

(339b) Agent-Based Modeling Analysis of Biomass Feedstock Production System Dynamics and Resilience

Yogendra Shastri[1], Luis Rodriguez[2], Alan Hansen[2] and K.C. Ting[2], (1)Department of Agricultural and Biological Engineering, University of Illinois at Urbana-Champaign, Energy Biosciences Institute, Urbana, IL, (2)Department of Agricultural and Biological Engineering, University of Illinois at Urbana-Champaign, Urbana, IL

An efficient and sustainable energy crop production system is essential for the success of the emerging biorefinery industry. However, there is currently a significant lack of understanding regarding the development and dynamics of the biomass feedstock market upstream of the biorefinery, which is a critical component of the value chain. This includes assessing the time-scale of its development as well as the impact of various engineering, technological, regulatory and social factors on the dynamics. Various uncertainties and disturbances can impact the functioning of this system, and understanding the resilience of the system in the presence of those is also important. The analysis is especially challenging due to the involvement of multiple stakeholders such as farmers, refinery, transportation companies and storage elevators that must participate and yet compete for a fraction of the profit.

The objective of this work is to conduct such as analysis using the complex adaptive systems theory approach. The work develops an agent-based model that incorporates various stakeholders (agents) within the feedstock production system. The model, developed in MATLAB®, uses an object-oriented programming approach by considering each agent-type as an independent class and a particular realization of that agent-type as an object (instance) of that class. The individual characteristics of an object are defined by a set of attributes. The behavior of each agent as well as its interaction with other agents is based on a set of rules (methods). Since the behavioral and social factors play an important role in the decisions made by individual agents, a mental model incorporating some of these factors is also developed for each agent. The set of decision rules are consequently based not only on the economic and technological rationale, but also on the unique mental models and cognitive processes. The learning/ adaptive behavior exhibited by the agents is also modeled which enables the simulation of emergent behavior of the system.

The model has previously been used to study the development of the biomass market under different scenarios. The focus of this work is on understanding the impact of various disturbances and uncertainties on the dynamics of the system, which enables us to study the resilience of the system. Different possible disturbances such as loss of productivity, price fluctuations and infrastructure breakdown will be modeled. Desirable but sudden changes such as the availability of an efficient technology will also be considered. Simulation studies incorporating these disturbances will be conducted to study their impact on the dynamics of the feedstock production system. Preliminary studies using this model have illustrated that the development of this sector is gradual, nonlinear and may take up to 15 years. The impact of the time of disturbance along this development trajectory will therefore be critical and will also be studied. Simulations for different types of energy crops such as switchgrass and Miscanthus will also identify whether the dynamics are a function of the energy crop. The work will also enable us to evaluate different system resilience measures appropriate for analyzing the feedstock production system.

See more of this Session: Sustainable Biomass Feedstock Production and Supply for the Emerging Biorefinery Industry
See more of this Group/Topical: Sustainable Engineering Forum

(641b) A Method for Rapid Screening of Catalysts for Hydrodeoxygenation of Biomass Intermediates

David K. Johnson, Biosciences Center, National Renewable Energy Laboratory, Golden, CO

The focus of biofuels development is shifting from alcohol fuels to advanced biofuels that would be drop-in replacements for petroleum-derived hydrocarbons. There is a wide range of intermediates that can be produced from biomass by chemical or biological routes, but virtually all require further deoxygenation before they can be used as fuel components. Potential routes for decreasing the oxygen content of biomass components include dehydration, hydrodeoxygenation and decarboxylation (possibly after controlled oxidation). Deoxygenation by HDO is preferred over dehydration and decarboxylation reactions, as it is likely to result in higher yields through conservation of carbon, retention of hydrogen, and the production of more energy dense products. Hydrodeoxygenation (HDO) is a hydrogenolysis process that removes oxygen by converting it to water, however, HDO is often performed at high temperatures and pressures that significantly increase the cost of the transformation. We need to develop HDO catalysts that more efficiently convert biomass-derived components into fuel products that are compatible with the existing fuel distribution infrastructure. The feedstocks for the chemical transformations can include sugars, lignins, bio-oils, and chemicals obtained by biological transformations of sugars and biomass. It is clear from previous research that in order to design commercially viable catalysts for HDO upgrading of biomass-derived intermediates, the catalyst system must (1) exhibit high activity under mild conditions (<200°C), (2) contain low-cost materials, (3) demonstrate high stability in an aqueous environment, and (4) have high selectivity for the desired products. To accomplish this, a microwave heated (to 250°C), pressurizable (to 250 psi), automated reaction system has been assembled to rapidly screen catalyst systems using model compounds. To allow automated operation, hydrogen is provided in-situ from a hydrogen donor material, such as formic acid, *i*-propanol or decalin. The rapid screening not only allows the rapid testing of catalysts, but also of reaction conditions, including temperature, reaction time, solvent, and hydrogen donor. The results from tests with model compounds such as acetovanillone and hydroxymethyl furfural, representing typical lignin and carbohydrate derived intermediates, with heterogeneous catalysts, will be described.

See more of this Session: Chemical and Catalytic Conversions and Processes for Renewable Feedstocks
See more of this Group/Topical: Sustainable Engineering Forum

(471c) Anode Assembly and Templating Strategies for Improving Efficiency and Versatility of Dye-Sensitized Solar Cells

Pisist Kumnorkaew, **Mark A. Snyder** and James F. Gilchrist, Department of Chemical Engineering, Lehigh University, Bethlehem, PA

The operation and efficiency of dye-sensitized solar cells (DSSCs) are inextricably linked to both light and charge transport/management. Efforts to realize higher overall device efficiencies are confounded, in part, by challenges associated with the multicomponent nature of DSSCs. Small incremental improvements in device efficiency since the pioneering work by Grätzel [1] nearly a decade ago motivate comprehensive optimization efforts spanning molecular engineering of novel dye sensitizers to materials engineering and component integration. In this talk, we will describe efforts aimed at engineering anode microstructure in order to control and manage light transport without sacrificing conductivity or critical dye sensitization. This is accomplished by embedding convex and concave microlenses at the interface of the conductive glass and porous, nanostructured titania anode via convective particle deposition and sacrificial templating, respectively. The result is anode-embedded, ordered arrays of colloidal particles or concave replica structures that lead to enhancements in measured device efficiencies by as much as 28%. The convective assembly strategy enables systematic variation of microlens properties (e.g., coordination, composition, size, location), and provides a handle for tuning anode morphology and even assembling anode structures on flexible, conductive substrates. We will describe various fundamental studies of how modes of photon capture and distribution can be facilitated by embedded microlenses, how such light management translates to improvements in cell efficiency, and how progress towards realization of more versatile flexible devices can be realized.

[1] O'Regan, Grätzel, *Nature*, 353 (1991).

See more of this Session: Nanostructured Materials for Dye-Sensitized Solar Cells
See more of this Group/Topical: International Congress on Energy 2011

(355f) Surface Plasmon-Enhanced Selective Molecular Sensing Using Unique Silver Nanoaggregates

Marimuthu Andiappan, Phillip Christopher and Suljo Linic, Department of Chemical Engineering, University of Michigan, Ann Arbor, Ann Arbor, MI

The selective identification of molecular species in a mixture of species is extremely important for the detection of explosives and biological warfare agent and the mechanistic study of chemical reactions. In general, viable techniques for chemical identification must provide a combination of exceptional sensitivity and molecular selectivity. Recent developments in nanoparticle synthesis and nanofabrication techniques, combined with an increased understanding of plasmonic properties of nanomaterials (mainly silver and gold), have greatly transformed surface-enhanced Raman spectroscopy (SERS) into one of the most promising techniques for highly sensitive molecular identification. The resonant creation of surface plasmons on noble metal nanoparticle surfaces and the subsequent enhancement of local electric fields results in enhanced Raman signals of many order of magnitude, and in some cases even allowing single molecule detection. However, application of SERS technique in the selective identification of trace quantities of targeted species in a mixture of species still remains a difficult task.

In this contribution, we demonstrate that aggregates comprised of different shapes of silver nanoparticles (nanocubes, nanowires and nanospheres) exhibit different enhancement for the C≡C stretching (2219.6 cm-1) and phenyl ring breathing (996.2 cm-1) vibrational modes in SERS of diphenylacetalene (DPA). UV-vis extinction spectra were collected concurrently with the SERS spectra, allowing us to develop a correlation between the nanoparticle aggregate plasmonic properties and the SERS spectra. The results show that the unique wavelength dependent plasmon resonance of Ag nanoparticles of different shapes can be utilized to independently control the enhancement of the two monitored vibrational modes. The results indicate that the synthesis of SERS substrates with tailored plasmonic properties allows a novel route towards targeted selective molecular detection.

See more of this Session: Catalytic, Environmental, and Industrial Gas Sensors
See more of this Group/Topical: Topical 9: Sensors

(288d) Successful Networking-Yes, Engineers Can Network Too!

Lori McDowell, Praxair, Inc, The Woodlands, TX

The author will discuss networking.

See more of this Session: Developing Leadership Skills
See more of this Group/Topical: Management Division

(62f) Modeling and Optimization of Algae Based Biodiesel Production

Soumya Yadala, Chemical Engineering, The University of Tulsa, Tulsa, OK and Selen Cremaschi, Department of Chemical Engineering, The University of Tulsa, Tulsa, OK

Processing of biodiesel from algae biomass has attracted much interest in recent times due to many reasons. Significant amongst them are global warming and scarcity in fossil fuels. It has been suggested that this third generation biofuel, a carbon neutral alternative, is capable of meeting the existing demand for transport fuels [1]. However, substantial challenges are encountered inderiving fuels from algae biomass. Primary amongst them is the relativelyhigher cost of cultivating and harvesting algae species. Due to this reason,currently, algae based biofuels cost much higher than the biofuels from firstgeneration and second generation biofuel feedstock [2].

This work focuses on modeling and optimization of algae based biodiesel production. Design parameters, for cultivation of algal species, at a geographical location are affected by that location's climatic conditions such as annual solar irradiance. Hence, geographical location, at which algae is cultivated, plays a vital role in the biodiesel production. Another significant factor to be considered for producing maximum amount of biodiesel is the selection of algal species. There are a variety of algal species present throughout the world, however, not all algal species are suitable for biodiesel production because of varying amounts of oil content and specific growth rates. Hence, selection of algal species is also an important factor in determining the cost of biodiesel production. There are a number of technologies for each process involved in the production of biodiesel from algae biomass but the question of which technology, for which species, and at which location, to attain maximum profits is still an open challenge to the researchers. Figure 1 presents the tasks that should be completed to produce algae-based biodiesel and options/technologies (which can be used to achieve these tasks) considered in this work. Using Figure 1, a disjunctive programming model is developed to determine the algal species, the location for the cultivation of algae, and the optimum route from growth of algae to production of biodiesel by selecting one technology for each process with the objective of minimum cost of biodiesel production. The resulting Mixed Integer Nonlinear Programming(MINLP) model is solved using GAMS 23.6.5/SBB solver. With the model parameters used, 8000 tons/year of biodiesel can be produced at a cost of about $153 per gallon for a period of ten years using the selected technologies. Accordingly, Tricornutum species was selected to grow at all the given locations by means of artificial growing technique-tubular photo bioreactor. The cultivated algae wereharvested through flocculation which was then concentrated using centrifugation.The algae biomass was dried by low pressure shelf drying followed by extraction of algal oil by expeller press and finally, the algal oil was sent to transesterification for biodiesel production.

REFERENCES

1. Y. Chisti, Biodiesel from microalgae, *Biotechnol. Adv.* **25** (2007), pp. 294-306.

2. Sheehan, J., Dunahay, T., Benemann, J., Roessler, P. A Look Backat the U.S. Department of Energy's Aquatic Species Program: Biodiesel fromAlgae; Close-Out Report. U.S. Department of Energy (1998).

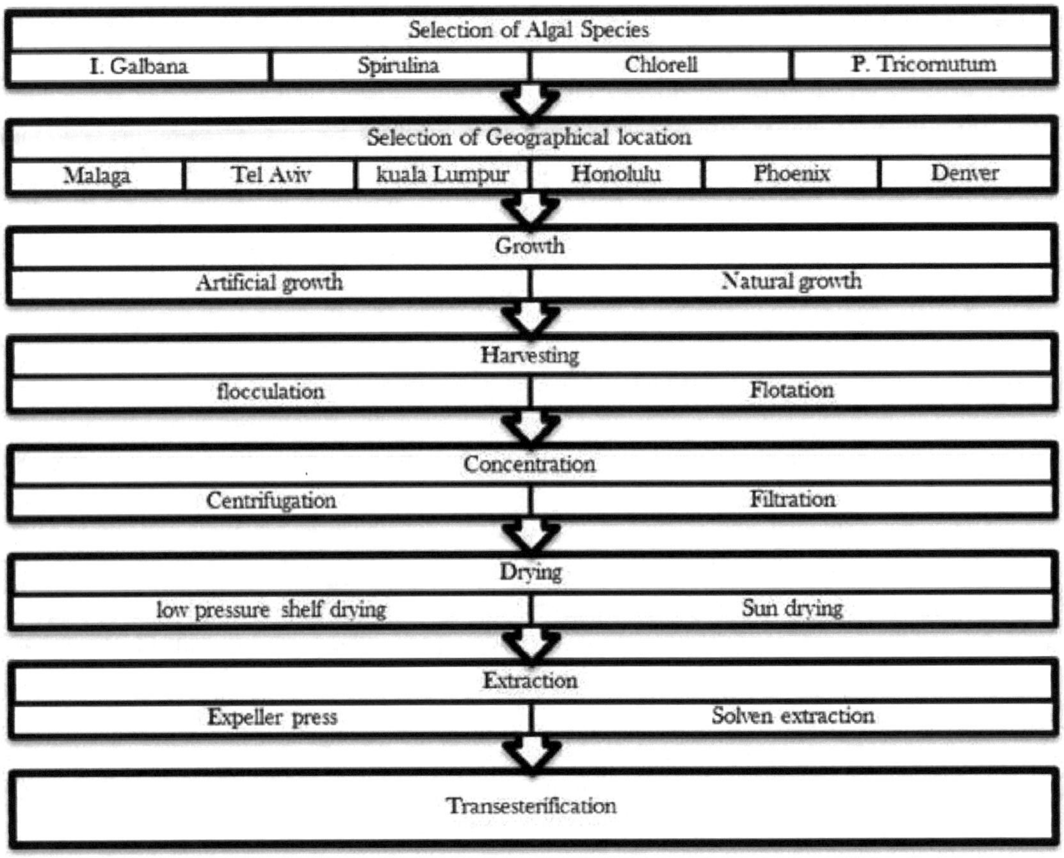

Figure 1. Tasks and corresponding options/technologies for biodiesel production from algae

See more of this Session: Advances In Algal Biorefineries I
See more of this Group/Topical: Sustainable Engineering Forum

(305f) An Integrated Framework for Design of Sustainable Chemical Process Based On Life Cycle Assessment

Masahiko Hirao and Yasunori Kikuchi, Department of Chemical System Engineering, The University of Tokyo, Tokyo, Japan

In this work, we propose a supporting framework for the design of chemical process integrating computer-aided process engineering (CAPE) and life cycle assessment (LCA) tools. LCA has been a useful tool for quantifying the environmental impacts from existing product life cycle and its application to process development enables us to design environmentally conscious product and process. During process development, however, implementation of LCA is difficult due to lack of inventory data. CAPE tools are used for simulation and evaluation of chemical processes, but have potential for the estimation of process inventories used in LCA. Thus, we modeled activities and information flows using integration definition for function modeling (IDEF0) method for the sustainable chemical process development in which CAPE and LCA tools are integrated. Applicability of the proposed activity model is demonstrated and verified by case studies on the process development of bioethanol-derived chemicals.

See more of this Session: Life Cycle Analysis of Renewable Feedstock-Based Processes and Products I
See more of this Group/Topical: Sustainable Engineering Forum

(10e) Selection and Characterization of Acid-Stable Cellulases Against Pretreated Oil Palm Empty Fruit Bunch From a Colombian Extreme Environment

Luis Miguel Medina[1], Luisa Cabezas[2], Isabella Bahamon[1], Laura Palma[1], Andrs Gonzlez[1] and Silvia Restrepo[2], (1)Chemical Engineering, Centro de Diseo de Productos y Procesos, Universidad de los Andes, Bogot, Colombia, (2)Biological Sciences, Laboratorio de Micologa y Fitopatologa (LAMFU), Universidad de los Andes, Bogot, Colombia

The rapid growth of the oil palm industry around the world generates environmental concerns, with a production of more than 28.0 million tons per year of oil palm around the world, generating a great amount of by-products and wastes, such as liquid effluent, gaseous emission, and solid waste materials. Nowadays, solid waste is used commonly as manure, nevertheless the composting process is reported to be complex[1]. Therefore this agro-industry should take into consideration the use of sustainable technology. A material balance of an oil palm plant yields, in terms of mass, for every Ton of fresh fruit processed 220 kg of oil palm empty fruit bunch (OPEFB), 125-130 kg of fiber, around 50 kg of shell and 195-200 kg of palm oil [2]. Studies on the lignocellulosic content of the OPEFB have shown that display a high cellulose and hemicellulose and low lignin content [2]. This feature allows to use this waste as a substrate to obtain simple sugars, which can be used as carbon source in different fermentation processes to enable the production of alcohols that can be used as fuels. The biodegradation of cellulose is carried out by three classes of enzymes: endoglucanase (carboxymethylcellulase), exogluconase (cellobiohydrolase) and β-glucosidase, which together with a preliminary treatment of cellulose, manage to break the glycosidic bonds to produce simple sugars [3]. However, these enzymes hydrolyze in different parts of the polymer so the relation between the substrate and the enzyme should not be extrapolated due to different lignocellulosic structure [4]. Taking all this into account the objective of this study was to explore the cellulolytic potential of endophytes from a Colombian extreme environment for the saccharification of oil palm empty fruit bunch. It was determined the cellulolytic capacity for the saccharification *in vitro* of the pretreated oil palm empty fruit bunch of one hundred Colombian native endophytes isolated from young leaves attacked by insects (with herbivory) as well with no symptom of attack (without herbivory) from *Espelieta grandiflora* and *E. corymbosa* plants from the Páramo Crúz Verde. The microorganisms were screened to asses which of them have cellulolytic capacity on solid agar with CMC as substrate followed by Congo red staining method [5]. The microorganisms that showed cellulolytic activity were grown on cellulase inducting medium, using pretreated OPEFB as substrate, the activity of the three main components of cellulase (CMCase, β-glucosidase and Cellobiohydrolase) was measured, and total cellulolytic activity (Filter paper activity. FPA), using the procedures described by Zhang *et. al (2009)*. Candidates obtained, (two entophytes), were partially purified through partial precipitation with ammonium sulfate (90%) [6] and chromatography using DAE-shepadex A-50 column previously equilibrated with sodium acetate buffer (pH 5, 0.05 M) and the fractions were withdrawn with a linear NaCl gradient, the fractions with cellulolytic activity were pooled and then concentrated by ultrafiltration with a centriprep (Amicon) having a 10 kDa molecular mass cut-off. The screenings allowed us to identify 28 endophytes with cellulolytic activity, and the best obtained were a *Penicilium sp.* and an *Aspergillus sp.* with maximum enzyme activities, CMCase, FPase, exogluconase and β-glucosidase, of 0.50-0.16, 0.035-0.032, 0.87-0.99 and 0.068-0.01 U/ml respectively. CMCase and the FPA activities of the partially purified cellulase was assessed at pHs ranging from 1 to 10, and temperatures ranging from 30 to 80 °C in order to find the optimum conditions for the cellulases. The cellulases from *Penicillium sp. and Aspergillus sp.* were stable at low pH which could be a valuable tool for hydrolysis of cellulose under acidic conditions.

References

1. Yusoff, S., *Renewable energy from palm oil - innovation on effective utilization of waste.* Journal of Cleaner Production, 2006. **14**(1): p. 87-93.

2. Umikalsom, M.S., et al., *The treatment of oil palm empty fruit bunch fibre for subsequent use as substrate for cellulase production by Chaetomium globosum Kunze.* Bioresource Technology, 1997. **62**(1-2): p. 1-9.

3. Sun, Y. and J. Cheng, *Hydrolysis of lignocellulosic materials for ethanol production, a review.* Bioresource Technology, 2002. **83**(1): p. 1-11.

4. Zhang, Y.H.P., M.E. Himmel, and J.R. Mielenz, *Outlook for cellulase improvement: Screening and selection strategies.* Biotechnology Advances, 2006. **24**: p. 452-481.

5. Kasana, R., et al., *A Rapid and Easy Method for the Detection of Microbial Cellulases on Agar Plates Using Gram's Iodine.* Current Microbiology, 2008. **57**(5): p. 503-507.

6. Scopes, R.K., *Protein purification principles and practice* Second ed. Spinger Advanced texts in chemistry, ed. C.R. Cantor. 1982: Springer-Verlag.

See more of this Session: Biological Conversions and Processes for Renewable Feedstocks I
See more of this Group/Topical: Sustainable Engineering Forum

(286b) Solar Thermochemical Recycling of CO2 Using ALD Deposited CoFe2O4 On Alumina Supports

Paul Lichty, Department of Chemical and Biological Engineering, University of Colorado, Boulder, CO, Xinhua Liang, Chemical and Biological Engineering, University of Colorado at Boulder, Boulder, CO, Carl Bingham, National Renewable Energy Laboratory, Golden, CO and Alan W. Weimer, Department of Chemical and Biological Engineering, University of Colorado at Boulder, Boulder, CO

Fuels production through metal oxide redox cycling can be completely renewable if reduction temperatures are achieved using concentrated solar energy. Many combinations of metal oxides have been investigated over the years and cobalt ferrites show much promise as they have good thermodynamic properties and high melting points. However, bulk ferrites degrade when cycled due to sintering of starting materials and resulting diffusion limitations. Recent work has found that coating the ferrites on rigid supports helps to reduce sintering. When using alumina as a support several Co/Fe/Al spinel species are formed which are capable of being cycled at lower temperatures. Cobalt ferrite has been coated on alumina nano powder and high surface area alumina shells. Large, high surface area polymer particles (~600μm, 43.5 m^2/g) were coated via atomic layer deposition (ALD) with several layers of alumina. The alumina layers were applied using TMA/water ALD chemistry. The particles were then heated to 800°C in air to burn out the polymer substrate. The alumina substrates were then coated with alternating layers of Fe_2O_3 and Co-O using ferrocene and cobaltocene ALD chemistries.

ICP results also show metal oxide mass loadings up to 27%. Thermo-chemical cycling was performed in a Netzch DIL 402 thermogravimetric analyzer with CO production measured using an in-situ Netzch QMS 403C mass spectrometer. Particles with alumina supports maintain cycling efficiency as compared to unsupported particles. Testing has been conducted on sun in a multi-tube reflective cavity reactor at the National Renewable Energy Laboratory in Golden, Colorado. Cycling results help to estimate reactor efficiencies.

See more of this Session: Concentrated Solar for Power Generation and Chemical Processing I
See more of this Group/Topical: International Congress on Energy 2011

(471a) Hydrothermal Fabrication of Ordered SnO2 Nanorod Arrays by Liquid Phase Conversion Process for Dye-Sensitized Solar Cells

Umang V. Desai and Di Gao, Chemical and Petroleum Engineering, University of Pittsburgh, Pittsburgh, PA

One-dimensional semiconductor materials have attracted significant research interest because of their unique properties for various technological applications. SnO_2, as a wide band-gap semiconductor material is a promising photoanode material for dye-sensitized solar cells because of its fast electron mobility and suitable conduction band-edge position. Crystalline, ordered SnO_2 nanorod arrays were successfully grown directly on a transparent conducting oxide using a hydrothermal approach. The fabrication procedure involved using an ordered ZnO nanowire array as a template and then converting it into SnO_2 nanorod array via a liquid-phase conversion process. Large-scale growth of uniform nanorod arrays, with high orientation consistency, was achieved through this low temperature hydrothermal process. Furthermore, the performance of SnO_2 nanorod arrays as photoanode in dye-sensitized solar cells was evaluated.

See more of this Session: Nanostructured Materials for Dye-Sensitized Solar Cells
See more of this Group/Topical: International Congress on Energy 2011

(258e) Coupled Fluid-Dynamics and Population-Balance Kinetic Models for Enzymatic Hydrolysis of Biomass

Michael A. Sprague[1], Jonathan Stickel[2] and Andrew Griggs[2], (1)Computational Science Center, National Renewable Energy Laboratory, Golden, CO, (2)National Bioenergy Center, National Renewable Energy Laboratory, Golden, CO

A population-balance kinetic model developed previously for the enzymatic hydrolysis of lignocellulosic biomass is extended to include fluid mixing and transport. The kinetic model is focused on the depolymerization of insoluble cellulose chains by an endoglucanase and an exoglucanase. Population-balance models combined with a continuous-distribution representation provide computational efficiency, which is critical when considering the addition of fluid-mixing effects. The resulting coupled equations are solved with a high-order open-source computational fluid dynamics (CFD) package. The CFD model is validated against experimental and analytical data for a simple vane mixer. Preliminary results examining the effects of fluid mixing on dilute enzymatic hydrolysis are presented and compared to existing experimental data. This work provides a foundation for future efforts to model pilot- and commercial-scale bioconversion-reactors.

See more of this Session: Reaction Kinetics and Transport Fundamentals for Biomass Conversion
See more of this Group/Topical: Sustainable Engineering Forum

(305e) Effect of Biomass Feedstock On the Sustainability Metrics of Fischer-Tropsch Fuels

Ignasi Palou-Rivera and M.Q. Wang, Argonne National Laboratory, Argonne, IL

The use of biomass as feedstock for the manufacture of Fischer-Tropsch fuels has significant effects on their sustainability metrics. A number of studies have looked at this biomass use [1,2], with an emphasis on techno-economic results. Some of the studies incorporate limited life-cycle analyses [1,2,3], especially considering green-house gas emissions. Our study is a complete Well-To-Wheels Life Cycle Analysis of biomass-fed Fischer-Tropsch liquid fuels (gasoline, diesel, and jet-fuel) performed using our GREET(Greenhouse gases, Regulated Emissions, and Energy use in Transportation) Model developed at Argonne national laboratory under DOE sponsorship. The presented LCA has a particular focus on the effects of several design and operational parameters:

- ***Biomass characteristics***: forestry and agricultural residues, humidity level, relative amount (when not used as only feedstock), etc.
- ***Gasification characteristics***: biomass co-gasification with coal, mix of biomass-derived syngas with other (e.g. coal- and natural gas-derived syngas), gasifier technologies, etc.
- ***Process characteristics***: CO_2 capture and sequestration (CCS) degree, process integration, FT synthesis characteristics and catalyst type, etc.

The LCA includes an encompassing sensitivity analysis of the effects of all these parameters on major sustainability metrics. The considered metrics include energy use (total, fossil, etc), green-house gas emissions (CO_2e GHG), and emissions of other criteria pollutants (NO_x, PM10, PM2.5, SO_x, VOC, CO)

[1] Liu, G.; Larson, E.D.; Williams, R.H.; Kreutz, T.G.; Guo, X. *Making Fischer-Tropsch Fuels and Electricity from Coal and Biomass: Performance and Cost Analysis*. Energy Fuels, 2011, 25, 415-437

[2] NETL/DOE (National Energy Technology Laboratory/U.S. Department of Energy), *Affordable, Low-Carbon Diesel Fuel from Domestic Coal and Biomass*, January 2009.

[3] Xie, X.; Wang, M.; Han, J. *Assessment of Fuel-Cycle Energy Use and Greenhouse Gas Emissions for Fischer-Tropsch Diesel from Coal and Cellulosic Biomass*. Environ. Sci. Technol., 2011, 45, 3047-3053.

See more of this Session: Life Cycle Analysis of Renewable Feedstock-Based Processes and Products I
See more of this Group/Topical: Sustainable Engineering Forum

(471e) Rapid Deposition of Titania Nanoparticles for Dye Solar Cells

Rocco Panella, B. Erik Ydstie and Dennis C. Prieve, Chemical Engineering, Carnegie Mellon University, Pittsburgh, PA

Dye-sensitized solar cells (DSSCs), while only having been designed about twenty years ago, are considered to be one of the most promising third-generation solar cell technologies among experts. One of the key features of DSSCs is a mesoporous, 10 micron layer of titania nanoparticles which acts as a high surface area scaffold for the application of a highly photoactive dye. The properties of this titania nanoparticle layer, such as its porosity, electrical interconnectivity, and resistance to cracking, are important for the performance of the solar cell as well as for their implications on large-scale manufacturing of this technology.

Using electrokinetics and fluid mechanics, we are designing a method to rapidly deposit titania nanoparticles from aqueous suspensions onto transparent conductive oxide (TCO) surfaces for use in DSSC anodes. Manipulating the pH and ionic strength of an aqueous solution makes it possible to control the surface potential of submerged oxide surfaces such as titania and doped tin oxide. By using pH values near 5.5, it is possible to create conditions where titania nanoparticles are stable and electrostatically opposed to each other in a dilute, aqueous suspension, but are attracted to tin oxide surfaces, including both tin-doped indium oxide (ITO) and fluorine-doped tin oxide (FTO). We can then use fluid flow such as that from a rotating disk or impinging jet to deliver nanoparticles adequately close to the surface such that they adhere via strong van der waals forces. For particles which are made in solution via oxidation and sol-gel processes, this technique offers the major advantage of never drying the particles, thus allowing them to remain un-aggregated until they deposit onto the surface.

We have shown that we can monitor particle deposition *in situ* by measuring the streaming potential of a surface using a rotating disk electrode apparatus. The measured differences in iso-electric point of titania nanoparticles and surfaces such as FTO and ITO shown in this work indicate that we should have control over the rate and extent of particle deposition. Currently, we have deposited titania nanoparticle layers which are several layers thick and consist of particles with diameters smaller than 10 nm; these tightly packed layers have applications as blocking layers to prevent unwanted shunts in dye solar cells. Improvements in the short-circuit current of DSSCs prepared using blocking layers applied in this fashion have been demonstrated. We also describe the potential for the scaling of a process based on fast, electrokinetic deposition to the size of specialty-coated glass production, which is recognized as a long-term goal for solar cell mass production.

See more of this Session: Nanostructured Materials for Dye-Sensitized Solar Cells
See more of this Group/Topical: International Congress on Energy 2011

(144d) Novel Electrochemical Sensor to Detect Heat and Sweat Condition Inside the Sockets of Prosthetics

Nathaniel J. Blasdel and Chelsea N. Monty, Department of Chemical and Biomolecular Engineering, The University of Akron, Akron, OH

Novel electrochemical sensor to detect heat and sweat conditioninside the sockets of prosthetics.

Themost current estimate states there are approximately 1 in 190 persons living inthe United States with major limb-loss while the rate of amputations completed increaseseach year. This necessitates an importance for understanding both quality oflife rated (QOLR) issues and options to remediate prevalent problems. There area number of important issues affecting people living with the loss of a limband the number one issue is ambulation or replacing the functionality of thelost limb. Throughout time we have developed many different styles ofprosthetics to fill this deficiency, while the socket style prosthetic hasevolved to be one of the most common solutions. These socket style prosthetics facilitatea need to replace a missing limb, but with this comes problems associated withthe use of the device. Proper fitting is important, along with proper care ofthe device and residual limb. These are issues that can be controlled by theamputee. One major issue that is uncontrollable for amputees is the combinationof heat and sweat in the socket. This is troublesome for a residual limb,because the socket can become hot and humid during even just regular use andcan cause a variety of dermatological conditions if proper care is not taken. Themajor contributors to heat and sweat inside the socket are personal activity,and socket and liner materials of construction. Socket and liner materials ofconstruction negatively affects the socket environment by inhibiting heattransfer away from the residual limb and just ten minutes of walking canincrease the average residual limb temperature by 1.7°C. A reduction in heattransfer causes sweat inside the socket, which can create a moist, abrasiveenvironment between the skin, sock or liners, and the prosthetic. The purposeof this presentation is to discuss the feasibility of monitoring heat and sweatby repurposing a well-known colorimetric chemistry for electrochemicaldetection. The sensing unit incorporates the reaction between silver chromateand chloride ions in sweat. This reaction changes color from brown to white andthis can be detected as a voltage change at an electrode surface. Thispresentation describes the design and fabrication methods of a novelelectrochemical sensor, the surrounding concepts for characterizing theelectrical properties of electrospun polymer thin films and their surface andfiber structures, and surface and encapsulated reaction kinetics.

See more of this Session: Biosensor Devices III
See more of this Group/Topical: Topical 9: Sensors

(305d) Pennycress-Derived Jet Fuel and Diesel LCA: Sustainable Advanced Biofuels

Jiqing Fan, Chemical Engineering, Michigan Technological University, Houghton, MI, David Shonnard, Michigan Technological University, Houghton, MI, Tom N. Kalnes, Research and Development, UOP LLC, Des Plaines, IL and Peter Johnsen, Arvens Technology Inc, Peoria, IL

Pennycress-derived jet fuel and diesel show promise as emerging biofuels that can help reduce dependence on petroleum imports, lower the trade deficit, and most importantly, lower greenhouse gas (GHG) emissions. Pennycress is a prolific seed producer with yields of more than 2000 pounds of per acre. The seeds contain 36% oil by weight, nearly twice the amount as soybeans. In the envisaged biofuel scenario, Pennycress would be grown between segments of a typical corn/soybean rotation, thus avoiding the concerns of a fuel crop competing with food crop production. As useful applications for pennycress meal develop, the oil is expected to become a sustainable alternative for advanced biofuels production.

The goal of this study is to determine the life cycle GHG emissions, cumulative energy demand, and fossil energy demand of pennycress-derived Hydroprocessed Renewable Jet (HRJ) fuel and Renewable Diesel (RD) as produced by the UOP Ecofining^{TM} process, and compute the GHG savings per MJ of fuel compared to petroleum-based jet fuels and diesel. The scope of this study encompasses the entire life cycle from pennycress cultivation and raw materials acquisition through the production and use of the fuels in vehicles and aircraft operations. The pennycress is grown in the Midwestern United States as a winter annual. After harvesting, the pennycress seeds are transported to a centralized processing facility where the oil is recovered and a de-oiled meal co-produced for use as a secondary energy source or animal feed. The oil is then transported to a hydroprocessing plant as a source of renewable feedstock for HRJ and RD production. Inventory data for pennycress cultivation, transportation, and oil recovery were provided by Arvens Technology Inc. Data for conversion of the pennycress oil to HRJ and RD were obtained from engineering design data, and was supplied by UOP. All the inventory data were assembled based on energy content of 1 MJ of final fuel product, which was the functional unit of this life cycle assessment (LCA) study. In the base case, energy allocation was applied to distribute environmental impacts among various products and co-products along the life cycle of each biofuel. The life cycle GHG emissions (excluding land use change) for RD and HRJ are 26.5 and 28.5 g CO_2 eq./MJ fuel, which represent savings relative to petroleum fuel counterparts of 72% and 69% respectively. Total energy consumption for each pennycress biofuels product is comparable to petroleum fuels with the exception that most of the energy is derived from renewable biomass as opposed to non renewable fossil. Scenario analyses were also conducted to determine response to model assumptions and data uncertainty, including allocation methodology, fertilizer application rate, source of H_2, and land use change. The results show that pennycress derived biofuels could qualify as advanced biofuels as defined by the Renewable Fuels Standard (RFS2).

(10d) Inhibition of ENZYMATIC Hydrolysis by Soluble Sugars ON MODEL CELLULOSE Thin Film USING QUARTZ CRYSTAL Microbalance

Hsin-Fen Li[1], Ravinder Garlapalli[1], Michael D. Flythe[2], Sue E. Nokes[3], Stephen E. Rankin[1] and Barbara L. Knutson[1], (1)Chemical and Materials Engineering, University of Kentucky, Lexington, KY, (2)Department of Animal and Food Sciences, University of Kentucky, Lexington, KY, (3)Department of Biosystems and Agricultural Engineering, University of Kentucky, Lexington, KY

INHIBITION OF ENZYMATIC HYDROLYSIS BY SOLUBLE SUGARS ON MODEL CELLULOSE THIN FILM USING QUARTZ CRYSTAL MICROBALANCE

Hsin-Fen Li [1], Ravinder Garlapalli[1]. Michael D. Flythe [2,3], Sue E. Nokes [4], Stephen E. Rankin[1], Barbara L. Knutson[1]

[1]University of Kentucky, Department of Chemical and Materials Engineering, Lexington, Kentucky 40506

[2]USDA, Agricultural Research Service, Forage-Animal Production Research Unit; Lexington, Kentucky 40546

[3]University of Kentucky, Department of Animal and Food Sciences, Lexington, Kentucky 40546

[4]University of Kentucky, Department of Biosystems and Agricultural Engineering, Lexington, Kentucky 40506

Soluble sugars produced from the hydrolysis of cellulose are one of the most promising feedstocks for production of fuels and commodity chemicals from renewable plant-based resources. The recalcitrance of cellulose to degradation is well documented and attributed to multiple sources, including high cellulose crystallinity and the hydrogen binding framework that is embedded in the matrix substances (lignin and hemicellulose). Cellulase is a cellulytic enzyme that possess a catalytic domain capable of hydrolyzing the beta-1, 4-glycosidic bonds in cellulose. However, during biomass conversion to soluble sugars, its hydrolytic efficiency is limited by the presence of degradation products including cellobiose. To study the inhibitory effect of cellobiose and the interaction of the inhibited enzyme and the cellulose substrate, the change in mass and surface properties of cellulose thin films were measured using Quartz Crystal Microbalance with Dissipation (QCM-D). The cellulose thin film was synthesized from thermochemically dissolved cellulose and was coated on a quartz crystal resonator. The mass of the spin-coated cellulose film was detected by the resonance frequency, and the cellulose surface coverage was observed by atomic force microscopy (AFM). Cellulase adsorption, activity, and inhibition on model cellulose thin films as were measured directly as changes to the piezoelectric properties of the cellulose thin films by QCM-D. At 5 g/L, cellobiose completely inhibits cellulase (0.5 %v/v) activity. The inhibited cellulases bind reversibly to the cellulose and reduce subsequent enzymatic hydrolysis rates by more than 50%. Significantly higher activity is observed when cellulase concentration is decreased from 1 %v/v to 0.1 %v/v, suggesting that cellulase acts progressively on the cellulose chain, and the hydrolysis is restricted by saturated enzyme conditions.

See more of this Session: Biological Conversions and Processes for Renewable Feedstocks I
See more of this Group/Topical: Sustainable Engineering Forum

(13e) A Dual-Stage Laminar Entrained Flow Reactor (LEFR) and Plug Flow Reactor (PFR) System for Studying Biomass Pyrolysis and Gasification

Mark W. Jarvis, Calvin Mukarakate, David Robichaud and Mark Nimlos, National Renewable Energy Laboratory, Golden, CO

An understanding of pyrolysis vapor cracking kinetics and mechanisms are important for developing approaches for reducing tars during gasification and ultimately reducing the cost of producing transportation fuels from gasification and fuel synthesis. In order to reduce tars using predictive design approaches such as Computation Fluid Dynamics (CFD), it is necessary to have robust chemical reaction models containing mechanism that lead to the formation of these species. Existing reaction models only contain limited, highly generalized mechanisms for the formation of tars and are not robust enough to be predictive. In particular, the initial steps of gasification are inadequately characterized. These steps are primarily cracking reactions that break down the pyrolysis vapor species into smaller molecules ultimately leading to syngas and tars. In this study, we have undertaken an experimental investigation of these types of cracking reactions. The solid phase vaporization reactions were separated from the gas phase cracking reactions by conducting experiments in a two-stage reactor. In the first stage, particles of biomass were pyrolyzed in a Laminar Entrained Flow Reactor (LEFR) at a low temperature (450°C) and in the second phase the vapors were cracked at variable residence times (0.2 to 1.5 s) and temperatures (430° to 680°C) in a Plug Flow Reactor (PFR). The tar species and other product gases were measured using a Molecular Beam Mass Spectrometer (MBMS). The data from the MBMS showed trends in the tar species concentrations as a function of residence time and temperature, which will allow the development of more accurate and robust chemical kinetics models of gasification. Eventually, these kinetic models will be used in CFD models of gasification in order to optimize it for reduced tar formation and increased syngas yield.

See more of this Session: Catalytic Conversion of Renewable Resources to Synthesis Gases and Pyrolysis Oils
See more of this Group/Topical: Sustainable Engineering Forum

(62c) Effects of Total Inorganic Carbon and Nitrogen Concentrations On Lipid Formation In Chlorella Vulgaris

Jinsoo Kim, Department of Chemical & Materials Engineering, University of Cincinnati, Cincinnati, OH, Joo-Youp Lee, Department of Chemical and Materials Engineering, University of Cincinnati, Cincinnati, OH and Kaniz F. Siddiqui, The Metropolitan Sewer District of Greater Cincinnati, Cincinnati, OH

Effects of totalinorganic carbon and nitrogen concentrations on lipid formation in *Chlorella vulgaris*

Jinsoo Kim[1], Joo-YoupLee[1], Kaniz F. Siddiqui[2]

[1] Chemical EngineeringProgram, University of Cincinnati, OH 45221-0012

[2] The Metropolitan SewerDistrict of Greater Cincinnati, Cincinnati, OH 45204-2002

Microalgae have recently received greatattention as a promising alternative bioenergy source because they can produce asubstantial amount of triacylglycerols (e.g. 20-50 % dry cell weight) and canbe readily converted into biodiesel. *Chlorella vulgaris* has been selected forthis growth kinetic study since it is one of the fastest growing microalgae andcontains a reasonable amount of triacylglycerols (e.g. 14-22 % dry cell weight). Dissolved carbon dioxide (CO_2(aq)) and bicarbonate ion (HCO_3^-) areessential inorganic carbon forms required for autotrophic algal growth, but fewstudies have been systematically conducted on the impact of inorganic carbon onalgal growth. In this respect, thegrowth kinetics of *Chlorella vulgaris*was previously investigated dependent on the total inorganic carbonconcentration (i. e., CO_2(aq) and HCO_3^-) As a result, it was observed that the growthkinetics was increased when the total inorganic carbon concentration wasincreased in the culture medium.

Under many nutrient limiting conditions,lipids are accumulated as a means of storage when energy (i.e. light) andcarbon (i.e. CO_2 for autotrophic growth) sources are available. However, many of these nutrients areessential for photosynthesis responsible for metabolic flux generation intolipids formation. Thesenutrient-deficient conditions decrease biomass growth rate, and offset thebenefit of lipid accumulation. Thegrowth of algal biomass such as *Chlorella vulgaris* is dependent on bicarbonate ion concentrationsavailable in the culture medium. Bicarbonateion also plays a critical role as a catalyst in the first step of fatty acidsbiosynthesis for the synthesis of malonyl-CoA.In this study, a change in total lipids content has been investigatedusing the Nile red stain method when the total inorganic carbon concentrationin culture medium is increased. This isexpected to provide fundamental understanding of the utilization of bicarbonateions for the design and operation of bioreactors.

Keywords: *Chlorella vulgaris*, total inorganic carbon, nitrogen deficiency,lipid content.

See more of this Session: Advances In Algal Biorefineries I
See more of this Group/Topical: Sustainable Engineering Forum

(254e) Highly Active Oxide Photocathode for Photoelectrochemical Water Reduction

Elijah Thimsen[1], Adriana Paracchino[2], Vincent Laporte[3], Kevin Sivula[2] and Michael Graetzel[2], (1) Energy, Environmental and Chemical Engineering, Washington University in St. Louis, St. Louis, MO, (2)Institute of Chemical Sciences and Engineering, Ecole Polytechnique Fdrale de Lausanne, Lausanne (VD), Switzerland, (3)Interdisciplinary Centre for Electron Microscopy, Ecole Polytechnique Fdrale de Lausanne, Lasuanne (VD), Switzerland

A clean and efficient way to overcome the limited supply of fossil fuels and the greenhouse effect is the production of hydrogen fuel from sunlight and water through the semiconductor/water junction of a photoelectrochemical (PEC) cell, where energy collection and water electrolysis are combined into a single semiconductor electrode. We present a highly active photocathode for solar H_2 production, consisting of electrodeposited cuprous oxide, which was protected against photocathodic decomposition in water by nanolayers of Al-doped zinc oxide and titanium oxide deposited by atomic layer deposition (ALD) and activated for hydrogen evolution with electrodeposited Pt nanoparticles. The roles of the different surface protection components were investigated, and in the best case electrodes showed photocurrents up to –7.6 mA cm^{-2} at a potential of 0 V vs. RHE (Reversible Hydrogen Electrode) at mild pH. The electrodes remained active after 1h of testing, cuprous oxide was found to be stable during the water reduction reaction and the Faradaic efficiency was estimated to be close to 100%.

See more of this Session: Process Development and Design for Solar-Grade Silicon Production
See more of this Group/Topical: International Congress on Energy 2011

(258a) Mechanistic Modeling of Simultaneous Enzymatic Saccharification of Cellulose and Xylan Using Continuous Distribution Kinetics

Ambarish Nag[1], Andrew J. Griggs[2], Jonathan J. Stickel[2], Michael A. Sprague[1] and James J. Lischeske[2], (1)Computational Science Center, National Renewable Energy Laboratory, Golden, CO, (2) National Bioenergy Center, National Renewable Energy Laboratory, Golden, CO

Cost effective production of biofuels and biochemicals often involves enzymatic saccharification, which has been the subject of intense research and development. In this work, we develop a mechanistically based kinetic model for the enzymatic hydrolysis of biomass composed of cellulose, xylan, and lignin. The prototypical biomass particle in our model has a cellulosic core that is partially covered by xylan and lignin that impede the access of glucanase enzymes to the cellulosic core. Our model hence incorporates both glucanase and xylanase enzymatic saccharification of surface xylan and the cellulosic core. For degradation of insoluble cellulose chains we consider the actions of endoglucanase (endo-β-1,4-glucanase,) and exoglucanase (exo-β-1,4-cellobiohydrolase). Likewise for insoluble xylan we consider the actions of endoxylanase (endo-β-1,4-xylanase) and exoxylanase (xylan β-1,4-xylosidase), the latter enzyme hydrolyzing soluble xylan oligomers as well. The enzyme system also includes beta-glucosidase for the hydrolysis of soluble glucan oligosaccharides. The kinetics for the depolymerization of cellulose and xylan chains are described using population-balance equations, which are capable of describing the evolution of the distribution of chain sizes during hydrolysis, without solving equations for all chemical species present in the reacting mixture. This approach affords computationally efficient simulations, enabling the use of the model for process design and optimization. In combination with experiments, the kinetic model is used to study the transformations of the cellulose and xylan chain-size distributions during hydrolysis by an enzyme cocktail of all the enzyme species.

See more of this Session: Reaction Kinetics and Transport Fundamentals for Biomass Conversion
See more of this Group/Topical: Sustainable Engineering Forum

(471f) Self-Consistent Field Modeling of Microstructure Formation In Fluorinated Block Ionic Liquids for Photovoltaic Cells

John B. McLaughlin[1], Sitaraman Krishnan[1], Lin Wu[1], Lalitha V. N. R. Ganapatibhotla[1], Xinli Jia[1], Dipankar Roy[2] and Jianping Zheng[2], (1)Chemical and Biomolecular Engineering, Clarkson University, Potsdam, NY, (2)Physics, Clarkson University, Potsdam, NY

Ionic liquids are commonly used in electrolytes for dye-sensitized solar cells. To prevent leakage from devices, and to simplify device manufacture, it is desirable that the liquids have large viscosities. However, ionic conductivity is inversely proportional to viscosity [1]. To achieve high ionic conductivity at low fluidity, a new fluorinated imidazolium iodide ionic liquid with a block microstructure was synthesized. The chemical structure of this ionic liquid consists of linear blocks of fluoralkyl and poly(ethylene glycol) (PEG) segments with an imidazolium cation that is connected to the PEG block. The salt also contains free iodide anions. The conductivity and viscosity of electrolytes prepared using this ionic liquid were characterized using electrochemical impedance spectroscopy and cone and plate viscometry, respectively, at different temperatures. Microphase segregation of the immiscible fluoroalkyl and PEG segments resulted in an electrolyte with low fluidity (at temperatures below about 90 °C). The relative concentrations of different chemical groups in the electrolyte were found to have a strong effect on experimentally determined molar conductivity. The variations in conductivity could be explained based on the predictions of a self-consistent field model (SCMF) [2,3] on the formation and nature of microstructures in the electrolytes. This talk will discuss the results of the SCMF simulations in the context of experimental ionic conductivity data.

References

1. Ganapatibhotla, L. V. N. R.; Zheng, J.; Roy, D.; Krishnan, S. PEGylated Imidazolium Ionic Liquid Electrolytes: Thermophysical and Electrochemical Properties. *Chem. Mater.* **2010**, *22*, 6347-6360.
2. Scheutjens, J.M.H.M.; Fleer, G.J. Statistical Theory of the Adsorption of Interacting Chain Molecules. 1. Partition Function, Segment Density Distribution, and Adsorption Isotherms. *J. Phys. Chem.* **1979**, *83*, 1619-1635.
3. Scheutjens, J.M.H.M.; Fleer, G.J. Statistical Theory of the Adsorption of Interacting Chain Molecules. 2. Train, Loop, and Tail Size Distribution. *J. Phys. Chem.* **1980**, *84*, 178-190.

See more of this Session: Nanostructured Materials for Dye-Sensitized Solar Cells
See more of this Group/Topical: International Congress on Energy 2011

(144c) In-Situ Analysis of Organics In Exhaled Breath Utilizing Acid Catalyst Membranes

Adam D. Worrall[1], Jonathan A. Bernstein[2] and Anastasios Angelopoulos[1], (1)School of Energy, Environmental, Biological, and Medical Engineering, University of Cincinnati, Cincinnati, OH, (2) Division of Immunology, University of Cincinnati Medical School, Cincinnati, OH

Measurement of acetone content in human breath has been suggested as a non-invasive method to the diagnosis of diabetes and/or ketoacidosis. Existing methods use time consuming, ex-situ analyses of breath condensate that limit broad, real-time application. In the present study, the diffusion and immobilization of chemically and optically active probe molecules into the heterogeneous domains of perfluorosulfonic acid (PSA) polymer membranes is investigated for direct, one-step gas phase acetone analysis utilizing a hand-held device. The membranes act as acid catalysts for the condensation reaction of acetone and the immobilized agents at temperatures above the acetone boiling point, yielding an optical response in the visible region of the electromagnetic spectrum that may be observed utilizing standard light emitting diodes (LED's). Partitioning of the immobilized concentration among the lipophilic, hydrophilic, and transitional domains within the PSA membrane is shown to control the sensitivity of the membrane to acetone exposure. The detection limit of this method is found to be less than 1 ppb and is thus comparable to that of more typical, but ex-situ GC/MS and HPLC analyses. Using this method, it is possible to observe the real-time exhaled acetone concentrations of diabetic and non-diabetic patients during periods of fasting as a means to correlate breath acetone levels to blood glucose concentrations and the extent of ketoacidosis in diabetic patients. Extensions of this approach to other organic molecules that can be used as biomarkers of human health in exhaled breath are discussed for the development of multi-sensor arrays capable of real-time medical diagnostics and environmental monitoring.

See more of this Session: Biosensor Devices III
See more of this Group/Topical: Topical 9: Sensors

(512d) Identifying & Transforming Local Best Practices to a Global Competence

Marcus A. Tennant, Yokogawa Corp. of America, Pleasant Prairie, WI

Often in global corporations, best practices emerge that enable a company to improve business performance at alocal facility or region. In this presentation we will look at how best practices emerge locally but are often ignoredin a global perspective due to their local or regional origin. Many of these best practices are developed "below theradar" and work very well for a local market; however small changes and the proper strategy for adaptation canturn the best practices into company-wide global competence. We will we will discuss methods that a companycan "dig out" these practices and find the critical success factors, discuss strategies of implementing these bestpractices in other regions and how they could be used to improve business performance globally. Finally, we willalso look at potential pitfalls and how to overcome obstacles in a global implementation effort.

See more of this Session: Challenges In Open Innovation and Growth
See more of this Group/Topical: Management Division

(339f) Improving the Quality of Spent Mushroom Substrate (SMS) As Feedstock for Biorefineries

Maria Nydia Ruiz-Felix, Chemical Engineering, Villanova University, Villanova, PA, Justin M. Yeash, Department of Chemical Engineering, Villanova University, Villanova, PA and Justinus A. Satrio, Chemical Engineering Department, Villanova University, Villanova, PA

In the current ecological climate there is a strong push to find a renewable energy resource to reduce the usage of fossil fuels which are depleting and are causing greenhouse gas emissions. Biorenewable materials, particularly lignocellulosic biomass, have received increased attention recently as the only carbon-containing resources to produce renewable energy, particularly liquid transportation fuels. Many biomass resources that have been explored in the past include wood chips, tree bark, corn stover, switch grass, bagasse, and other organic wastes. Some of these sources are byproducts of commercial agriculture and some provide opportunities for biomass to energy agribusiness, where the crop is grown just for energy harvesting. At Villanova University, we are interested in research on biomass to energy conversions because the school is situated in the proximity of some major agricultural areas in the Mid-Atlantic region, like Lancaster County in Pennsylvania, which produces numerous biorenewable materials that can be used for energy production. Another important region near Villanova University is Chester County, which is known to be "the mushroom growing capital of the world". Currently, Chester County accounts for around 37% of mushroom production in the U.S.

Mushroom growing produces a large amount of waste product called spent mushroom substrate (SMS). After the mushroom fungi grow and fruit the nutrients in the substrate are no longer favorable for mushroom growth and the substrate is considered waste. SMS poses a unique opportunity as a renewable energy resource since it is available in very large amounts, especially in Chester County and the surrounding areas in Pennsylvania. SMS is typically land applied in nearby agricultural fields or sold as compost. These avenues provide little benefit to the mushroom growing companies and are unsustainable due to the methane released through the decomposition of the substrate.

As is, SMS is not the most ideal feedstock for biorefining due to its high moisture and ash/inorganic materials content. Our research is to explore how to improve the quality of SMS to make it an ideal feedstock for biorefining by using torrefaction as a pretreatment method. Torrefaction is the process by which a biomass source is heated in the absence of oxygen to between 200 and 300°C for residence times ranging from 15 to 60 minutes. In this process the biomass will lose 30% of its mass, but only lose 10% of its initial energy which results in significant energy densification. This energy densification is important for biomass feedstocks which may require transportation from the source to further processing facilities. Additionally, torrefaction helps to make biomass properties uniform which enables processing facilities to expend less energy adjusting to various feedstock properties. One of the crucial steps in the conversion of biomass to energy is making sure that the biomass characteristics are within optimal specifications for pyrolysis and further thermal degradation.

To be reported are the results of the torrefaction of SMS at various process temperatures and lengths of processing time. Properties being evaluated are heating values, ash content, and elemental chemical composition. These properties will be analyzed to determine the effects that torrefaction and cleaning by mechanical separation have on the quality of the SMS. Results from fast pyrolysis of torrefied/cleaned SMS compared to untreated SMS by using pyro-GC/MS will be reported as well. Finally, assessment on the sustainability of the torrefaction and cleaning by mechanical separation process for improving the quality of SMS will be presented.

See more of this Session: Sustainable Biomass Feedstock Production and Supply for the Emerging Biorefinery Industry
See more of this Group/Topical: Sustainable Engineering Forum

(144g) Single Molecule Detection of Nitric Oxide Enabled by d(AT)$_{15}$ DNA Adsorbed to near Infrared Fluorescent Single-Walled Carbon Nanotubes

Jingqing Zhang[1], Ardemis A. Boghossian[1], Paul W. Barone[1], Alina Rwei[1], Jong-Ho Kim[1], Dahua Lin[2], Daniel A. Heller[1], Andrew J. Hilmer[1], Nitish Nair[1], Nigel F. Reuel[1] and Michael S. Strano[1], (1) Department of Chemical Engineering, Massachusetts Institute of Technology, Cambridge, MA, (2) Department of Computer Science, Massachusetts Institute of Technology, Cambridge, MA

We report the selective detection of single nitric oxide (NO) molecules using a specific DNA sequence of d(AT)$_{15}$ oligonucleotides, adsorbed to an array of near-infrared fluorescent semiconducting single-walled carbon nanotubes (AT$_{15}$-SWNT). While SWNT suspended with eight other variant DNA sequences show fluorescence quenching or enhancement from analytes such as dopamine, NADH, L-ascorbic acid, and riboflavin, d(AT)$_{15}$ imparts SWNT with a distinct selectivity toward NO. In contrast, the electrostatically neutral polyvinyl alcohol enables no response to nitric oxide, but exhibits fluorescent enhancement to other molecules in the tested library. For AT$_{15}$-SWNT, a stepwise fluorescence decrease is observed when the nanotubes are exposed to NO, reporting the dynamics of single-molecule NO adsorption via SWNT exciton quenching. We describe these quenching traces using a birth-and-death Markov model, and the maximum likelihood estimator of adsorption and desorption rates of NO is derived. Applying the method to simulated traces indicates that the resulting error in the estimated rate constants is less than 5% under our experimental conditions, allowing for calibration using a series of NO concentrations. As expected, the adsorption rate is found to be linearly proportional to NO concentration, and the intrinsic single-site NO adsorption rate constant is 0.001 s^{-1} μM NO^{-1}. The ability to detect nitric oxide quantitatively at the single-molecule level may find applications in new cellular assays for the study of nitric oxide carcinogenesis and chemical signaling, as well as medical diagnostics for inflammation.

Zhang, J.; Boghossian, A. A.; Barone, P. W.; Rwei, A.; Kim, J.-H.; Lin, D.; Heller, D. A.; Hilmer, A. J.; Nair, N.; Reuel, N. F.; Strano, M. S. Journal of the American Chemical Society 2010, 133, 567.

See more of this Session: Biosensor Devices III
See more of this Group/Topical: Topical 9: Sensors

(231f) Integrations of Biosurfactant Production Into Advanced Biorefineries

William Colonna[1], **Mustafa E. Marti**[1], Michelle Pynn[2], Gabriel Reznik[2], Kevin Jarrell[2], Buddhi Lamsal[3] and Charles E. Glatz[1], (1)Chemical and Biological Engineering, Iowa State University, Ames, IA, (2)Modular Genetics, Inc., Woburn, MA, (3)Food Science and Human Nutrition, Iowa State University, Ames, IA

Several classes of biosurfactants are produced by microorganisms. One such surfactant is FA-Glu, which is related to the natural biosurfactant, surfactin. FA-Glu consists of a long-chain β-hydroxy fatty acid coupled to glutamic acid. It is produced by a strain of *Bacillus subtilis* genetically-engineered by Modular, Inc. It has several potential applications including use as a dispersant for oil spills, where biodegradability is desirable.

Economic production and eventual application will be favored if substrate costs can be reduced and low-cost recovery methods developed. Several such opportunities arise in advanced biorefineries. Product value would be further enhanced if such production could achieve favored "green" rating by avoiding solvents other than water and simple alcohols and effect emulsification and low toxicity could be demonstrated. This report focus on production considerations.

Production of FA-Glu by fermentation on glucose is compared to results using lower cost substrates available in advanced biorefineries, including fiber components of soy and glycerol. FA-Glu yields from soybean hulls (SBH) ranged from 29-51% of that from glucose on a substrate weight basis.. When SBH polysaccharides were hydrolyzed (~65%) by pre-treatment with carbohydrases, FA-Glu levels were only ~21-37% of those from glucose. The cells consumed the enzymatically-released sugars; however, very little of this was converted into FA-Glu. Addition of glucose to the enzymatic SBH hydrolysate improved FA-Glu yields, but only slightly. The low FA-Glu yields from SBH could be due to inhibitors released from the hulls by the enzymes or inhibitory substances in the enzyme preparations.

Glycerol, a by-product of biodiesel production, proved to be a suitable carbon source for FA-Glu production. Cells grown on 2% w/v glycerol produced ~25% more FA-Glu than cells grown on 2% glucose. Glycerol is abundant and low in cost. Its use as a feedstock for production of microbial biosurfactants would provide a means for efficient utilization of soy by-products and improving the ag economy.

The FA-Glu was purified by collecting the foam from the aerated broth, clarification by centrifugation or filtration, and adsorption and elution from an anion exchange column to a purity of 97.9%. Final product color can be improved by liquid-liquid extraction, but this may compromixe "green" status. A precipitation based alternative process is being examined as a means of eliminating solvent use.

(641e) Effects of Reaction Conditions On the Acid-Catalyzed Hydrolysis of Miscanthus Dissolved In An Ionic Liquid

Sean J. Dee and Alexis T. Bell, Chemical and Biomolecular Engineering, University of California - Berkeley, Berkeley, CA

Miscanthus, a perennial grass, is an attractive biofuel feedstock compared with corn and poplar due to its high energy yield per hectare and low agricultural inputs such as water and soil nutrients. Recent studies have shown the three primary components of miscanthus, cellulose, hemicellulose, and lignin, can be dissolved in ionic liquids (ILs) at levels approaching 20 wt%. An interesting question is then whether the cellulosic and hemicellulosic components of the dissolved miscanthus can be hydrolyzed to sugars in high yield. While acid-catalyzed hydrolysis of cellulose and hemicellulose has been studied extensively in ILs, the conversion of biomass is not as well understood. Recent literature has claimed that effective chemical conversions of biomass in ILs requires either thermochemical delignification which can destroy both hemicellulose and cellulose, or multistage processing that is inefficient and makes IL recycling energy intensive. This study investigated the acid-catalyzed hydrolysis of miscanthus dissolved in the ionic liquid 1-ethyl-3-methylimidazolium chloride ([Emim][Cl]). The effects of varying temperature, acid concentration, miscanthus concentration, and the gradual addition of water were undertaken. The similarities and differences between the rate of hydrolysis of microcrystalline cellulose (Avicel) and amorphous hemicellulose (Xylan from birchwood) compared to the cellulosic and hemicellulosic components of miscanthus were determined. The influence of miscanthus delignification on the hydrolysis rate was also explored. The present work has demonstrated that 90% biopolymer conversion from miscanthus could be achieved, with 93% selectivity to saccharine products in a single stage hydrolysis without pretreatment.

See more of this Session: Chemical and Catalytic Conversions and Processes for Renewable Feedstocks
See more of this Group/Topical: Sustainable Engineering Forum

(308d) Metal Oxide Nanowire Sensor System for Hazardous Gas Detection

Xiaopeng Li[1], Yang Shu[1], JungHwan Cho[2], Ying Wang[3], Yu Lei[4], Pradeep U. Kurup[2], Hongwei Sun[5] and **Zhiyong Gu**[1], (1)Chemical Engineering, University of Massachusetts Lowell, Lowell, MA, (2)Civil and Environmental Engineering, University of Massachusetts Lowell, Lowell, MA, (3)Chemical, Materials, and Biomolecular Engineering, University of Connecticut, Storrs, CT, (4)Chemical Engineering, University of Connecticut, Storrs, CT, (5)Mechanical Engineering, University of Massachusetts Lowell, Lowell, MA

Nanowires, featuring various metal oxide sensing materials such as copper oxide and doped tin oxides, have been successfully synthesized. Sensor devices were designed and fabricated to integrate nanowires as a functional component, where dielectrophoretic assembly was applied to orient and align the nanowires. Hazardous gases, mainly organic vapors, were tested with these nanowire sensor devices at different concentration levels. Copper oxide nanowire sensor was also chosen to detect hydrogen sulfide due to their unique reaction. Four types of doped tin oxide nanowire sensors were consolidated into an array in order to resolve the issue of poor selectivity. Both dynamic and static methods were utilized to obtain the response kinetics and sensitivity data for analysis. Sensing performance was characterized and principle component analysis (PCA) was conducted to further reveal the sensor capability of classifying the gas analytes.

See more of this Session: Micro and Nanofabricated Sensors II
See more of this Group/Topical: Topical 9: Sensors

(13g) Low Temperature Gasification of Impregnated Biomass Feedstock

Foster Agblevor, Biological Engineering, Utah State University, Logan, UT, Ofei Mante, Biological Systsems Engineering, Virginia Polytechnic Institute and State University, Blacksburg, VA, Allen Aradi, Afton Chemical Inc, Richmond, VA and Tze-Chi Jao, Afton Chemical, Richmond, VA

Hybrid poplar biomass feedstocks were impregnated with catalysts and gasified at three different temperatures in a bubbling fluidized bed reactor. The impregnated biomass were gasified in a nitrogen atmosphere at 400 °C, 600° C, and 800 °C. The gas yield increased with increasing temperature as expected and the char and liquid yields decreased with increasing temperature. Te catalytic effect was observed at 400 C when the impregnated gasification was compared with the control that did not contain any catalyst. At 400 C, the impregnated sample had a gas yield of 51wt% compared to 41 wt% for the non-catalytic gasification at 450 C. The char yield was 11 wt% compared to 14 wt% for the non-catalytic run at 450 C. The char yield decreased from 11 wt% to 6.5 wt% when the gasification temperature was increased from 400 to 800 C. Although there was 88 wt% gas yield at 800 C, the thermal effect was very strong and the catalyst was not effective in cracking the lignin degradation products. Thus, this catalyst needs considerable improvement to effectively crack the lignin fraction of the biomass pyrolysis products.

See more of this Session: Catalytic Conversion of Renewable Resources to Synthesis Gases and Pyrolysis Oils
See more of this Group/Topical: Sustainable Engineering Forum

(339e) Coupled Optimization and Simulation for Multi-Biomass Source-to-Biorefinery Supply Chain Modeling and Analysis

William Faulkner[1], Joseph Amundson[1], Dr. Fazleena Badurdeen[2] and Dr. Jeffrey Seay[3], (1)University of Kentucky, Lexington, KY, (2)Institute for Sustainable Manufacturing, University of Kentucky, Lexington, KY, (3)Chemical and Materials Engineering, University of Kentucky, Lexington, KY

As cost associated with fossil fuels rise due to various environmental, social, and economic factors, it has become prudent to examine the possibility of exploiting alternative sources of raw materials for the production of energy and chemicals. Biomass, of various forms, has been proposed as an option to fill this need for alternative raw materials. Several problems with biomass as an alternative to fossil fuel derived materials instantly arise: low bulk energy density, the resulting high transport costs, seasonal variability of biomass supply, and the risks inherent to relying on a largely agriculturally based system are several examples. To begin to realize energy and chemical production with biomass as the feedstock on an economic scale, supply chain optimization and a robust means of simulating variation over long periods of time is critical.

Due to a relatively low energy density of the given biomass feedstocks, the majority of the biorefinery supply chain costs occur in the process of raw material transportation. By optimizing this aspect of the biorefinery supply chain, given the plant and potential feedstock source locations, the cost associated with raw material transportation can be minimized. In order to examine the potential longevity and general robustness of the output from the optimization model, simulation is a valuable tool. In this way, the effects of perturbations on the various feed stock supplies due to weather variability, crop yield variability, or other factors can be examined and analyzed.

This presentation focuses on the development of a coupled system comprised of an optimization model for the upstream portion of biorefinery supply chains in Western Kentucky and a discrete-event simulation program. The optimization model minimizes the transportation costs associated with moving biomass feedstock from centralized storage facilities to the biorefinery location. Multiple feedstocks available in the region under consideration including chicken litter, forest residue, and corn stover were taken into consideration. Various alternative shipping methods as well as a centralized versus a distributed production facility layout are considered as well. The discrete-event simulation model allows for the examination of the robustness of the determined optimal supply chain design. This combination of optimization and simulation for the examination of multi-biomass source-to-biorefinery supply chains could potentially serve as a template for future work in biorefinery supply chain model development and decision-making.

See more of this Session: Sustainable Biomass Feedstock Production and Supply for the Emerging

(76d) A Lipobead Microfluidic Array for the Detection of Bacterial Pathogens

Xiaoxiao Chen[1], Miao Lu[1] and Charles Maldarelli[2], (1)Chemical Engineering, The City College of New York, The School of Engineering, New York, NY, (2)Levich Institute and Chemical Engineering, The City College of New York, New York, NY

The spread of bacterial infections through food and water borne pathways still poses a significant threat to the public health. This presentation focuses on the detection and identification of bacterial infections which cause disease by the secretion of bacterial proteins – exotoxins. Exotoxin-mediated bacterial infections include cholera, pertussis (whopping cough), tetanus and botulism, among many others. The spread of these bacteria through the food and water supply is of much greater prevalence in poorer countries lacking in infrastructures for the protection of the public health.

Rapid detection and identification of the infection is the first line of defense in preventing an epidemic, and this task is made more difficult by the large number of bacterial infections under surveillance. The goal of this research is to develop a new ultraminiaturized platform for multiplexed pathogen sensing which should allow for the rapid screening of an analyte sample for the presence and identification of a bacterial pathogen. We outline a bead-based microarray/microfluidic platform which can be used for the detection of multiple pathogens, and, because of the microfluidic miniaturization, requires much smaller sample volumes than immunoadsorbent assays. In this platform the pathogen sensing elements are micron-sized beads, 10-50 microns in diameter, which host probe molecules which conjugate to target pathogens potentially present in an analyte sample. These beads are arrayed in a microfluidic cell by entrapping the beads hydrodynamically in a microfluidic obstacle courses consisting of an array of traps. The traps are "V" shaped enclosures, which are arranged in a staggered configuration. Each trap contains an opening to allow flow through the trap in order to capture a single microbead. The array consists of different bead sets, with each set hosting on its surface a particular probe which binds to only one exotoxin. The sets are arrayed sequentially into the trapping array so that the probe on the surface of the bead in each trap is indexed.

The probe capture molecules used by the array are the natural glycolipid membrane receptors which the exotoxins use to target and attach to cells before infecting them. As such, these receptors bind the toxins with high selectivity and affinity. The receptors are displayed on the bead surface by first sequestering them in supported phospholipids bilayers which are formed around silica particles. Sequestering the receptor molecules in the lipid membrane allows them to retain their native binding ability. After the trapping array is filled, the analyte solution of potential targets can be flowed through the array, the binding events identified by fluorescently labeling the target and examining the array for fluorescing microbeads locations.

Experiments are presented on the detection of the cholera toxin using the glyolipid receptor GM_1 which binds to the toxin. A platform consisting of two bead sets, one containing the GM_1 and one control lipobead without a receptor are sequentially arrayed and the fluorescently labeled cholera toxin is streamed through the cell. Fluorescence mesurements indicate the exclusive binding of the toxin to the GM_1 lipobeads, demonstrating the use of the array for the detection of the toxin.

See more of this Session: Biosensor Devices II
See more of this Group/Topical: Topical 9: Sensors

(258c) Observing and Modeling Cellulosic Substrate Depolymerization by Commercial Enzyme Cocktails Using Confocal Fluorescence Microscopy

Jeremy S. Luterbacher[1], Jose M. Moran-Mirabal[2] and Larry P. Walker[2], (1)Department of Chemical and Biomolecular Engineering, Cornell University, Ithaca, NY, (2)Department of Biological and Environmental Engineering, Cornell University, Ithaca, NY

Understanding and modeling the depolymerization mechanisms of cellulosic and lignocellulosic substrates by cellulase cocktails is a critical step towards further optimizing the production of monosaccharides from biomass. The Spezyme CP cellulase cocktail combined with the Novo 188 b-glucosidase blend was used to depolymerize bacterial microcrystalline cellulose (BMCC), filter paper cellulose and untreated and H_2O-CO_2 pretreated lignocellulosic substrates (hardwood and switchgrass), which were immobilized on a glass surface. The enzyme mixture was supplemented with a small fraction of fluorescently labeled *T. Reseii* CBH I, which served as a reporter to track cellulase diffusion into and binding onto the internal physical structure of the cellulosic substrate. All reported experiments were conducted at 50°C, the optimal temperature for maximum hydrolytic activity of the enzyme cocktail. BMCC was observed throughout degradation by labeling it with a fluorescent dye. Native autofluorescence was used in a similar way to track filter paper and lignocellulosic substrates. This method allowed us to measure the binding of cellulases *in situ* and follow the temporal morphological changes of cellulose and lignocellulosic substrates during their depolymerization by a commercial cellulase mixture.

In the case of BMCC, a kinetic model was developed and fitted to fluorescence intensity data obtained through image processing. This model was successfully used to predict the bulk sugar concentrations that were liberated over time from BMCC in a separate experiment and measured by liquid chromatography. A more complete mass transfer and kinetic model is currently under development for the more complex substrates (filter paper and lignocellulosic biomass).

See more of this Session: Reaction Kinetics and Transport Fundamentals for Biomass Conversion
See more of this Group/Topical: Sustainable Engineering Forum

(641d) Catalytic Oxidative Dehydration of Butanol Isomers: 1-Butanol, 2-Butanol and Isobutanol

Ivan C. Lee, US Army Research Laboratory, Adelphi, MD

The catalytic production of economically important four-carbon olefins was achieved through catalytic oxidative dehydration of 1-butanol, 2-butanol, and isobutanol using a millisecond contact time reactor. Both alumina foam and rhodium-alumina foam catalysts convert these four-carbon alcohols into four-carbon olefins, with contact time equal to 25 ms over a wide range of equivalence ratios (f) from 0.176 to 6.338 (or C/O_{air} ratio from 0.0659 to 2.11) . The cracking of the carbon backbone was found to be minimal. A mixture of butene isomers was obtained due to the isomerization of the carbocation intermediates. The degree of this isomerization was determined by analyzing the ratios of butenes/(total olefins), 1-butene/2-butene, cis-2-butene/trans-2-butene, and isobutene/(linear butenes). In addition, a reaction mechanism of the catalytic oxidative dehydration of butanol in short contact times was proposed. In this mechanism, the C-O bond of the majority of butanol was broken to form olefinic species on the alumina surface. These surface olefinic species would undergo successive oxidation to form CO and/or CO_2, depending on the availability of oxygen.

See more of this Session: Chemical and Catalytic Conversions and Processes for Renewable Feedstocks
See more of this Group/Topical: Sustainable Engineering Forum

(34b) The Legal System As a Driver for Innovation In Energy Technology

Richard D. Jordan, IP, Dickinson Wright PLLC, Washington, DC and Mary Ternes, McAfee& Taft, Oklahoma City, OK

This presentation will provide an overview of how the legal system acts as a primary driver for innovation in energy related technology, considerations for funding from public and private sources, and considerations for determining how to best protect intellectual property developed in the innovation effort. The presentation will be made by invited policy and legal experts. These experts will discuss policy and procedural considerations in which governmental agencies, such as the Environmental Protection Agency, "intervene" for the good of society to require industry to provide "cleaner" technologies. These legal drivers in turn cause the industry to invest a great amount of funds to innovate in order to meet the new legal requirements. Panel experts will discuss: 1) a few of the policy issues that are considered by governmental institutions in implementing the new laws and regulations, 2) government grants or funds that are made available to stimulate innovation in such a capital intensive market, and 3) considerations on how to protect valuable intellectual property that will be created in getting the new innovations to market.

See more of this Session: Law, Risk, and Ethics of Technology and Project Management
See more of this Group/Topical: Management Division

(231b) Concentration of a Solids-Free Lignocellulose Hydrolysate for Very High Gravity Fermentation: Inhibition Effects

Steven J. Schneiderman, Todd J. Menkhaus and Patrick C. Gilcrease, Chemical and Biological Engineering, South Dakota School of Mines and Technology, Rapid City, SD

A major problem in the efficient conversion of lignocellulose hydrolysates to ethanol is low ethanol concentrations in the fermentation broth due to low initial sugar concentrations. This is undesirable because low ethanol concentrations lead to increased separation costs downstream. Distillation is energy intensive, and constitutes more than 50% of the total energy consumption of a typical ethanol plant. An obvious solution is to start with a high initial sugars concentration, which is referred to as a very high gravity (VHG) fermentation. By increasing the glucose concentration in the hydrolysate from 40 to 100 g/l, the distillation energy consumption would be reduced by 60%. In the first step of our integrated process, polyelectrolyte flocculants facilitate the efficient removal of biomass solids from a pine wood hydrolysate. Once the hydrolysate is solids-free, an evaporation or reverse osmosis step can be readily used to remove water and concentrate the fermentable sugars. VHG fermentations present some challenges of their own, including higher energy costs associated with concentrating the sugars, simultaneous concentration of some or all of the fermentation inhibitors, and osmotic stresses on the fermentation organism at higher substrate concentrations. Because some of the inhibition effects are synergistic, this could become more problematic at VHG conditions.

The goal of this study was to investigate the potential problems associated with fermenting a concentrated lignocellulose hydrolysate, and to determine operating conditions that maximize productivity and minimize costs for the combined sugars concentration-fermentation-distillation process. A batch culture design of experiments was used to quantify the inhibition effects of five common inhibitors on the productivity of our *S. cerevisiae* D_5A strain. The inhibitors studied were glucose, ethanol, furfural, HMF, and acetic acid at concentration ranges of 25 – 300 g/l, 0 – 150 g/l, 0 – 5 g/l, 0 – 5 g/l and 0 – 10 g/l, respectively. Our experimental results were used to regress kinetic model parameters, and define an acceptable operating window for VHG fermentation. The hydrolysate concentration step (evaporation/reverse osmosis) was then tailored to minimize total separation costs (sugars concentration and ethanol distillation) while maintaining a high level of ethanol productivity. Finally, the fermentation of a concentrated pine hydrolysate at optimum conditions was demonstrated at the bench scale.

See more of this Session: Integrated Processes for Biochemical Conversion of Renewable Feedstocks to Fuels and Chemicals I

See more of this Group/Topical: Sustainable Engineering Forum

(231c) Separate Processing of Hemicellulose and Cellulose to Optimize Ethanol Production From Corn Stover

Alexandre Chapeaux[1], Nancy Dowe[1] and Daniel J. Schell[2], (1)National Bioenergy Center, National Renewable Energy Laboratory, Golden, CO, (2)National Renewable Energy Laboratory, Golden, CO

Separate processing ofhemicellulose and cellulose to optimize ethanol production from corn stover

Alexandre Chapeaux, NancyDowe, Daniel J. Schell

Using a series of thermochemical (pretreatment)and biochemical steps, lignocellulosic biomass is converted to sugars that subsequentlycan be made into fuels and chemicals. One such process uses heat and dilute sulfuric acid to hydrolyzes the hemicellulosic fractionof the biomass to monomeric and oligomeric sugars.The removal of hemicellulose and to a lesser extent lignin from the biomass significantlyimproves the performance of the next step in the process, enzymatic hydrolysisof cellulose to glucose. However, enzymatic cellulose hydrolysis can beperformed in the presence or absence of the hemicellulosic sugars producedduring pretreatment. It is well known that removing the hemicellulose sugarsand other compounds produced during pretreatment improves enzymatic celluloseconversion yields. In previous work, we assessed the potential of a processthat removes the hemicellulosic sugars prior to enzymatic cellulose hydrolysis byexploring the interaction between biomass solids loading, yields, and processcosts. However, this process is more expensive because of the additional costfor solid-liquid separation equipment required to remove th-richsugar liquor from the cellulosic solids.

In this study, the goal was to deepen ourunderstanding of the advantages and disadvantages of the separatehemicellulose/cellulose process. The feedstock was dilute-acid-pretreated cornstover produced in a 200 kg dry biomass/d continuous horizontal reactor. Thepretreated slurry was separated into a liquor stream (hemicellulosic sugars)and a washed cellulosic solid stream. Enzymatic cellulose hydrolysis was investigatedat four enzyme loadings (10, 20, 30 and 40 mg protein/g cellulose) and atinsoluble solids concentrations from 9% to 21% (w/w). The resulting glucose wasfermented to ethanol using several yeasts and a glucose-xylose co-fermentingbacterium, *Zymomonas mobilis* 8b. The sugars in the liquorstream were fermented to ethanol with three strains of co-fermenting *Z. mobilis* after dilution of the streamwith water to lower sugar concentrations. Finally, the overall performance ofthe process was assessed after partitioning water between the two phases in away that optimizes overall ethanol yields. We then compared these results to aprocess configuration in which the streams are not separated. In the latterprocess, enzymatic cellulose hydrolysis is performed in the presence of thebackground hemicellulosic sugars produced during pretreatment of the biomass.

See more of this Session: Integrated Processes for Biochemical Conversion of Renewable Feedstocks to

(231e) Fischer Tropsch Synthesis Via Biomass Derived Synthesis Gas

Syed Ali Gardezi, Chemical Engineering, University of South Florida, Tampa, FL, Babu Joseph, Chemical and Biomedical Engineering, Clean Energy Research Center, University of South Florida, Tampa, FL and John T. Wolan, University of South Florida, Tampa, FL

Due to the recent surge in energy costs and uncertainties in fuel prices, there is significant interest in seeking alternative sources of energy particularly from renewable such as biomass. Indirect liquefaction of biomass (via the thermo-chemical conversion to syngas followed by liquefaction via Fischer-Tropsch Synthesis or FTS) offers a commercially viable route for meeting the challenge of producing renewable fungible liquid fuels. Apart from biomass, FTS is also the critical step for converting natural gas and coal to liquid fuels.

A FTS technology has been developed for producing middle distillates (diesel and aviation fuel) by using pine chips as the main feedstock. In this "two step" process, pine chips were subjected to an entrained flow gasification process developed by Pearson et al. [1]. The resulting tar free gas is comprised mainly of carbon monoxide, hydrogen, carbon dioxide, methane and some organic contaminants such as benzene, toluene, and naphthalene along with substantial amount of water. For further liquefaction this gas was cleansed using multiple adsorbents. An inline "Shaw Moisture Meter" was used to continuously monitor the moisture content of the feed to the liquefaction process.

For the production of liquid fuels, eggshell cobalt catalysts were used. Cobalt catalysts have several advantages e.g. they produce linear paraffinic hydrocarbons and minimize undesirable water gas shift reaction. This eggshell design is a well known strategy for producing a narrow hydrocarbon distribution. In this regard previous research work conducted by Iglesia et al. was also consulted [2]. A bench scale fixed bed reactor was used for the conversion of syngas. Modeling of the reactor startup and of adjustment of the reactor packing ensured successful operation resulting in a high conversion and narrow distribution of hydrocarbons in the range of diesel and aviation fuel.

References :

[1] Pearson et al., U.S. Patent 4,857,076, issued Aug 15, 1989.

[2] E. Iglesia, S. L. Soled, J. E. Baumgartner, S. C. Reyes, *J. Catal.* **153**, 108 (1995).

See more of this Session: Integrated Processes for Biochemical Conversion of Renewable Feedstocks to Fuels and Chemicals I
See more of this Group/Topical: Sustainable Engineering Forum

(11c) Surface Immobilization of DNA Aptamers for Biosensing and Protein Interaction Analysis

Xiaojuan Zhang and Vamsi K. Yadavalli, Chemical and Life Science Engineering, Virginia Commonwealth University, Richmond, VA

Molecular recognition between short synthetic oligonucleotide (DNA/RNA) sequences known as aptamers and their targets is increasingly being used for applications such as biodiagnostics and biosensors. To utilize aptamers as molecular recognition agents in biosensors and biodiagnostics, it is important to develop strategies for reliable immobilization of aptamers so that they retain their biophysical characteristics and binding abilities. Here we present quartz crystal microbalance (QCM) measurements and atomic force microscope (AFM)-based force spectroscopy studies to evaluate aptasensors. Gold surfaces were modified with mixed self assembled monolayers (SAMs) of aptamer and oligoethylene glycol (OEG) thiols (HS-C11-(EG)nOH) to impart resistance to nonspecific protein adsorption. By affinity analysis, we show that short OEG thiols have less impact on aptamer accessibility than longer chain thiols. Immunoglobulin E and vascular endothelial growth factor (VEGF) are presented as target proteins in these experiments. Binding forces obtained by these strategies are similar, demonstrating that the biophysical properties of the aptamer on the sensors are independent from the immobilization strategy. The results present mixed SAMs with aptamers and co-adsorbents as a versatile strategy for aptamer sensor platforms including ultrasensitive biosensor design. It is envisioned that this technology will enable insight into the fundamental biophysical interactions between aptamers and their target proteins for the development of sensitive sensor arrays and diagnostics.

See more of this Session: Biosensor Devices I
See more of this Group/Topical: Topical 9: Sensors

(144b) Handheld, Low-Cost Integrated CMOS Biomolecular Sensors

Stephen A. Chapman[1], Alex Pai[2] and Ali Hajimiri[2], (1)California Institute of Technology, Pasadena, CA, (2)Electrical Engineering, California Institute of Technology, Pasadena, CA

We are building our biomolecular sensor to perform as a Point-of-Care (POC) diagnostic. Our sensor would be suitable for many potential applications, such as detecting lethal agents in biological/chemical warfare, physiological sensing of molecular species, and diagnosing pathogens in epidemic disease outbreaks. The key technical issues for POC devices in these applications are hand-held portability, wireless integration for communication of data, high sensitivity, battery-level power consumption, fast-sensing process, and most importantly a low price-tag [3]. The resonance-based magnetic particle sensor scheme we have developed addresses these concerns, and is scalable, compatible with inexpensive bulk CMOS process, can be integrated seamlessly with wireless technology, and provides a true paradigm shift in the upcoming POC microarray market, making it possible to make extremely low-cost, self-contained sensors for a variety of applications.

See more of this Session: Biosensor Devices III
See more of this Group/Topical: Topical 9: Sensors

(258b) Kinetic Modeling of Cellulose Hydrolysis with First Order Inactivation of Adsorbed Cellulases

Zhuoliang Ye and R. Eric Berson, Chemical Engineering, University of Louisville, Louisville, KY

The rate of glucose release during enzymatic hydrolysis of cellulose slows down significantly as the reaction proceeds, and the mechanism is not yet clear. It is, therefore, desirable to develop a kinetic model that accurately describes the enzymatic hydrolysis reaction. In this work, a model based on Langmuir binding kinetics that also incorporates first order inactivation of adsorbed cellulase was developed that accurately predicts rates and yields during the hydrolysis reaction for two different cellulose substrates. Substrate concentrations of 20g/L and less were tested in order to remove any potential effects due to product inhibition or mass transfer limitations. A key result of the modeling suggests that Vmax decreases over time due to deactivation of the adsorbed enzyme. Rate constants associated with product formation and enzyme inactivation from our model yield are comparable to reported values elsewhere. The model predicted results within 10% of experimental results for both Solka Floc and Sigmacell substrates. Key model parameters were similar for the two substrates: the apparent hydrolysis rate (k_2) and inactivation rate (k_f) were 33 min^{-1} and 0.38 h^{-1}, respectively, comparable to values reported elsewhere. The rate constants were within 5% for the two different substrates, further strengthening the value of the model and implying that a common but yet undefined rate-limiting step associated with loss of enzyme activity likely exists in the pathway of cellulose hydrolysis. It is proposed that this rate-limiting step may be either dissociation or isomerization of the enzyme.

See more of this Session: Reaction Kinetics and Transport Fundamentals for Biomass Conversion
See more of this Group/Topical: Sustainable Engineering Forum

(13c) Multi-Scale Study On the Pyrolysis of Sustainable Biomass Feedstock

Jessica D. Murillo, Chemistry, Environmental Sciences, Tennessee Technological University, Cookeville, TN, Joseph J. Biernacki, Chemical Engineering, Tennessee Technological University, Cookeville, TN and C. Pat Bagley, College of Agricultural and Human Sciences, Tennessee Technological University, Cookeville, TN

Multi-Scale Study on the Pyrolysis of Sustainable Biomass Feedstock

Jessica D. Murillo

Department of Chemistry, Environmental Sciences, Tennessee Technological University, Cookeville, TN 38505

Joseph J. Biernacki

Department of Chemical Engineering, Tennessee Technological University, Cookeville, TN 38505

C. Pat Bagley

College of Agricultural and Human Sciences, Tennessee Technological University, Cookeville, TN 38505

Pyrolysis of lignocellulosic material for the production of biofuels and value-added products offers a short- to long-term renewable energy option to help curb fossil fuel use, gain energy independence, and mitigate global climate change. The main components of lignocellulsic material are cellulose, hemicellulose and lignin; however, there are many issues with using such a complex chemical feedstock for the production of renewable fuels. During pyrolysis the main components degrade via primary and secondary reactions, producing a spectrum of products, including volatiles, char, and bio-oil. To gain a more complete understanding of the pyrolytic behavior of lignocellulosic biomass, a deeper knowledge of the morphology and chemical structure is required.

Preliminary meso-scale experimental results from the pyrolysis of tall fescue hay, sorghum-sudangrass, Bermuda grass, switchgrass, and alfalfa hay, have paved the way for study at microscopic lengths. The physical and chemical characteristics of each grass hay was studied using X-ray diffraction, mid-infrared spectroscopy (MIR) and electron scanning microscopy to determine the microstructure and to quantify the content of lignin, cellulose, and hemicelluloses. Comparison studies on the reaction kinetics of grass samples and isolated pseudo-components were performed using thermogravimetric analysis (TGA).

Improving the sustainability of biofuels through modification of the biomass production subsystem: Application to corn and lignocellulosic ethanol

Concerns about dwindling fossil fuel reserves and climate change, energy independence, and depletion of ecosystem goods and services have driven the development of alternative technologies such as biomass-based fuels (biofuels). Although biofuels may be a step towards alleviating the aforementioned concerns due to the renewable nature of the biomass feedstock, there are several drawbacks associated with them. Food crop-based fuels such as corn ethanol, which comprise the majority of the current biofuel mix in the US, require large amounts of fossil-based inputs purchased from the economy in order to grow the biomass. These inputs, such as fertilizers, pesticides, and fossil fuels greatly diminish the renewable nature of the plant itself and force the production of biofuels to be reliant on non-renewable fossil-based inputs. In addition to reliance on non-renewable inputs, biofuels are heavily reliant on ecosystem goods and services. Current large-scale farming practices result in the runoff of nutrients from fertilizers into rivers and streams, which can overwhelm the carrying capacity of the nitrogen (N) and phosphorous (P) cycles of these aquatic ecosystems with devastating results. Freshwater for irrigation is needed if natural precipitation is not sufficient. Topsoil, a vital local non-renewable resource, is also depleted through erosion in traditional monocultural row cropping. Ecosystem goods and services are also indirectly consumed through the life cycle of the inputs used in farming.

This research aims to improve the sustainability of biofuels, specifically corn ethanol, by modifying the corn farming system to use less purchased inputs and consume less ecosystem goods and services both directly and indirectly through the reduction of fossil-based inputs needed from the economy. The following design changes will be considered:

1) No-till/conservation tillage. This will reduce soil erosion and require less labor and fuel for preparing soil for seeding. Application of more herbicides may be necessary as tilling helps remove weeds.
2) Planting winter cover crops in the growing offseason, also known as "double cropping". This reduces bare soil exposure in the offseason, reducing erosion. Also, offseason weed growth can be reduced requiring less herbicide when the corn is planted. Depending on the species planted, soil nutrients can also be rebuilt in the offseason. However, additional labor, energy, and other inputs will be needed for planting and harvesting the cover crop.
3) Intercropping other species amounts corn during the growing season. This is often referred to as a polyculture. This can reduce soil exposure and hence reduce erosion and reduce fertilizer consumption due to fixation of nutrients by the intercropped species. However, separation of corn from the intercropped species may require additional inputs.
4) Constructed wetlands and/or riparian buffers for nutrient runoff capture. This can prevent nutrients from entering rivers and streams and reduce eutrophication.

5) Using treated sludge from wastewater treatment facilities (biosolids) for fertilizer, displacing the need for fossil-based artificial fertilizers. However, there is a limited supply of biosolids and may not be enough for large-scale application. Also, biosolids may be more labor and energy intensive to apply.
6) Fossil fuel substitution, i.e., replacing partially or fully the fossil fuels used in the farming system with biofuels to reduce the reliance on non-renewable inputs. Any disadvantage of biofuels will be compounded when fed back into the production of more biofuels.

The impact of these design changes will be quantified both directly and indirectly in the life cycle through the use of Ecologically Based Life Cycle Assessment (Eco-LCA) to determine the overall effect of the changes on the sustainability of corn ethanol. This LCA method will quantify traditional LCA impacts, such as emissions and energy consumption, but will also quantify the multitude of ecological goods and services consumed throughout the life cycle. This ecological accounting is vital to ensure the sustainability of any system since everything relies directly or indirectly on ecosystem services. Additionally, the resulting corn farming design is an example of the design of a technological-ecological network, where both technological and ecological variables are considered in the design of the overall system. Such systems are more sustainable than systems that use technological variables only, since technology alone cannot lead to sustainability due to second law constraints.

Although these design changes may improve the sustainability of corn ethanol, the scaling up of biofuel production can compete with food and lead to indirect land use change where native ecosystems such as forests are converted to farmland for biomass production, which may create a net greenhouse gas increase. Therefore, it is worth considering other biomass sources that can be produced on marginal lands. The case of producing ethanol from a low input, high diversity (LIHD) lignocellulosic grassland on degraded farmland will also be presented. LIHD biomass requires much fewer inputs than most monocultural biomasses and may be a sustainable source of liquid fuel. LIHD ethanol will also be compared with food crop-based ethanol as well as other cellulosic ethanol sources.

(235f) Nanofabricated Chemical Sensors Using Aerosol Jet Printing Method

Ting Zhang, Zheng Cui and Rui Liu, Suzhou Institute of Nanotech and Nanobionics, Chinese Academy of Sciences, Suzhou, China

One-dimensional (1-D) nanostructures including semiconducting single-walled carbon nanotubes (SWNTs) have been demonstrated as good candidates for ultra-sensitive chemical sensors because of the high surface-to-volume ratio of nanostructures and their unique electronic conductance. However, the assembly of SWNTs across the microfabricated electrodes to form miniaturized sensors is always challenging and time-consuming with current methods, including guided CVD growth, polar molecular patterning, atomic force microscopy (AFM) manipulation, etc. The recent developments in printed electronics are expected to find an alternating way to fabricate cheap sensors by precisely printing well dispersed nanostructures on substrates. Among the printing techniques, the aerosol jet printing is one of the most promising techniques for making large area, inexpensive nanoelectronic components. It is a room temperature operated, and non-contact process which can be used to deposit a wide variety of materials onto a wide variety of substrates.

In this research, we demonstrate a cost effective, high-throughput and manufacturable printing method to fast fabricate semiconducting SWNTs based chemical sensors. Using aerosol jet printing method, Percolating networks of semiconducting SWNTs inks can be printed onto rigid or flexible substrates. The electrical properties of sensors were characterized with Field Effect Transistor (FET) configuration. The thin film transistors made by printing have demonstrated low operating voltage, good reliability and the printing process can be performed in air, in contrast to current organic thin-film transistors which have to be made in glove box in a controlled environment. Preliminary sensing results of printed flexible nanoelectronic sensors showed good sensitivity and fast response time to NH_3 and NO_2 at room temperature operation. The printing technology allows quick design and fabrication of electronic circuits, avoiding conventional semiconductor fabrication processes, which is a promising step toward the batch manufacturing of miniaturized sensor devices with low cost and extensive analytical capabilities.------------------

See more of this Session: Micro and Nanofabricated Sensors I
See more of this Group/Topical: Topical 9: Sensors

(231d) Ethanol and Furfural Production From Corn Stover Using Hybrid Fractionation Process with Zinc Chloride and Simultaneous Saccharification and Fermentation (SSF)

Chang Geun Yoo[1], **Tae Hyun Kim**[1] and Monlin Kuo[2], (1)Agricultural and Biosystems Engineering, Iowa State University, Ames, IA, (2)Department of Natural Resources Ecology and Management, Iowa State University, Ames, IA

Two-stage hybrid fractionation process was investigated to produce cellulosic ethanol and furfural from corn stover. Zinc chloride ($ZnCl_2$) was used to selectively solubilize hemicellulose in the first stage. During the second stage, the remaining treated solids were converted into ethanol using commercial cellulase and *Saccharomyces cerevisiae* or recombinant *Escherichia coli*, KO11.

Hybrid fractionation process recovered >90% of glucan, ~90% of xylan, and ~75% of lignin at optimal reaction conditions: 1st stage: 5% acidified $ZnCl_2$, 7.5 ml/min, 150 °C (10 min) and 170 °C (10 min); 2nd stage: Simultaneous saccharification and fermentation (SSF). Furfural yield from hemicellulose hydrolysates was 58 %. SSF of $ZnCl_2$ treated solid with *Saccharomyces cerevisiae* resulted in 69-98 % of theoretical maximum ethanol yields based on glucan present in the treated solids. After fermentation, solid residues contained mostly lignin. Based on total lignin in untreated corn stover, the lignin recovery yield was ~75%.

See more of this Session: Integrated Processes for Biochemical Conversion of Renewable Feedstocks to Fuels and Chemicals I
See more of this Group/Topical: Sustainable Engineering Forum

(254a) Solar Grade Silicon Production In Fluidized Bed Reactor

Juan Du, Soham Dutta and B. Erik Ydstie, Department of Chemical Engineering, Carnegie Mellon University, Pittsburgh, PA

In this work, we present a multi-scale modeling approach for solar grade silicon production in fluidized bed reactor (FBR) by thermal decomposition of silane. The FBR process has been commercialized by several companies and promises to deliver poly-silicon at reasonable and stable price. Silane decomposes to form hydrogen and silicon and seed particles grow by heterogeneous chemical vapor deposition and by scavenging silicon powder produced in homogeneous gas phase reaction. The subsequent particle growth process is represented by a discretized version of population balance. The model uses ordinary differential and algebraic equations to track particle movement through discrete size intervals to simulate changes in the size distribution. The dynamics of gas phase is modeled Computational Fluid Dynamics (CFD) model. The CFD module shows the hydrodynamics via momentum balance, mass and heat transfer between different phases. The volume fraction of phase, concentration of each components and temperature of the bed is generated and are imported into the population balance. The particle size distribution calculated by population balance is used as the input of the CFD module and thus the complex interplay between particulate phase and continuous phase is captured by integrating those two modules.

Based on the multi-scale model, a novel method is proposed to derive the global stability condition to guarantee the stable operation of FBR. The resulting analytical stability condition will be verified by numerical simulations. The particulate process in FBR is complex and typically has very few measurements, inventory control is a simple method for control of complex systems and thus has potential for industrial application. We apply inventory control strategy to control particle size distribution to improve the performance of the silicon production process. Due to the lack of knowledge of the reaction coefficient, which is used in population balance to describe the mass transfer rate from gas phase to dispersed phase, an on-line estimator based on adaptive mechanism will be developed to track experimental behavior more accurately and in addition to enhance robustness of the inventory control system.

See more of this Session: Process Development and Design for Solar-Grade Silicon Production
See more of this Group/Topical: International Congress on Energy 2011

(62a) Optimization of a Louisiana Native Algal Co-Culture for Biofuel Production

Rong Bai[1], Maria Teresa Gutierrez-Wing[2], **Michael G. Benton**[1] and Kelly A. Rusch[2], (1)Cain Department of Chemical Engineering, Louisiana State University, Baton Rouge, LA, (2)Civil and Environmental Engineering, Louisiana State University, Baton Rouge, LA

Microalgae have been reported to possess significantly higher lipid productivity per unit area than traditional oil crops, a benefit for biofuel production. Microalgal strains such as *Nannochloropsis sp., Botryococcus braunii,* and *Chlorella vulgaris* have been reported to contain high lipid/oil content. However, these results are mostly at the lab scale. In large-scale, commercial culture systems, tolerance to local environmental conditions and invasive algal species significantly affect lipid/oil production. Therefore, native microalgal strains with high lipid productivity are likely better options for large-scale biofuel production.

A native fresh water microalgae strain from Louisiana was investigated. This work measures the impacts of irradiance and nitrate levels on total lipid yield and the fatty acid profile of this strain. The culture was harvested in the late exponential-stationary phase to maximize the lipid content, and Soxhlet extraction with chloroform/ methanol (2:1, v/v) was used for total lipid content evaluation. Microwaves and nanoparticle assisted Folch's extraction methods have been used to optimize lipid extraction. The fatty acid profiles of neutral lipids were analyzed by gas chromatography combined with mass spectrometry.

The results indicated the lipid yield and content at 400 and 800 mmol $m^{-2}s^{-1}$ was the highest and no significant differences (32.13 and 32.76 mg $L^{-1}d^{-1}$, 31% and 30%) between the two treatment levels were observed. Results also indicate that although the lower nitrate level increased lipid content, the increase did not compensate for a decrease in growth rate. The fatty acid profiles indicate the major components are 16- and 18-carbon fatty acids, and ~35% are saturated fatty acids. The fatty acid profiles do not vary significantly with the irradiance and nitrate levels.

See more of this Session: Advances In Algal Biorefineries I
See more of this Group/Topical: Sustainable Engineering Forum

(144f) Label-Free, Single Protein Detection On a near Infrared Fluorescent Single-Walled Carbon Nanotube/Protein Microarray

Jin-Ho Ahn[1], Jong-Ho Kim[2], Jingqing Zhang[2], Nigel Reuel[2] and Michael Strano[2], (1)Chemical Engineering, Massachusetts Institute of Technology, Cambridge, MA, (2)Massachusetts Institute of Technology, Cambridge, MA

Excessive sample volumes continue to be a major limitation in the analysis of protein-protein interactions, motivating the search for label-free detection methods of greater sensitivity. Herein, we report the first chemical approach for selective protein recognition using fluorescent SWNT enabling label-free microarrays capable of single protein detection. Hexahistidine-tagged capture proteins directly expressed by cell-free synthesis on SWNT/chitosan microarray are bound to a Ni^{2+} chelated by Nα,Nα-bis(carboxymethyl)-L-lysine grafted to chitosan surrounding the SWNT. The Ni^{2+} acts as a proximity quencher with the Ni^{2+}/SWNT distance altered upon docking of analyte proteins. This ability to discern single protein binding events decreases the apparent detection limit from 100 nM, for the ensemble average, to 10 pM for an observation time of 600 sec. This first use of cell-free synthesis to functionalize a nanosensor extends this method to a virtually infinite number of capture proteins. To demonstrate this, the SWNT microarrays are used to analyze a network of 1156 protein-protein interactions in the staurosporine-induced apoptosis of SH-SY5Y cells, confirming literature predictions.

See more of this Session: Biosensor Devices III
See more of this Group/Topical: Topical 9: Sensors

(62b) Controls of Microalgal Biomass and Lipid Production In Municipal Wastewater-Fed Bioreactors

Belinda S.M. Sturm[1], Val H. Smith[2], Marie-Odile Fortier[1], Edward Peltier[1] and Frank Jerry deNoyelles[3], (1)Civil, Environmental & Architectural Engineering, University of Kansas, Lawrence, KS, (2)Ecology and Evolutionary Biology, University of Kansas, Lawrence, (3)Kansas Biological Survey, Lawrence

The use of wastewater as a nutrient source for algal growth as a biodiesel feedstock could increase the viability of this process and reduce wastewater discharge of N and P compounds. In this study, four algal bioreactors were fed continuously with municipal wastewater effluent over a six month period; these were monitored weekly to determine algal biomass and lipid production, nutrient removal, and carbon and nutrient availability. Algal biomass production ranged from 0.78 to 15.9 g dry weight m^{-2} d^{-1} over the experimental period. Algal reactors removed 19% of dissolved nitrogen and 43% of dissolved phosphorus from wastewater effluent. A stoichiometric analysis of particulates revealed that algal biomass had far lower C:P ratios (67:1) and N:P ratios (6:1), but higher C:N ratios (17:1), than Redfield values. These data suggest that the algae produced were rich in phosphorus, and that N was likely to be growth-limiting in these systems. Numerous studies also suggest that algal biomass can be enhanced by providing CO_2 in excess of normal atmospheric concentrations, and that algal biofuel production can contribute to CO_2 sequestration efforts. In order to optimize nutrient removal in wastewater-fed algal reactors, and to determine the potential for CO_2 sequestration, future reactor engineering research should continue to incorporate the expertise of ecological stoichiometry.

See more of this Session: Advances In Algal Biorefineries I
See more of this Group/Topical: Sustainable Engineering Forum

(76b) All-Optical Microfluidic Biosensor

Edward P. Furlani[1], Natalia M. Litchinitser[2] and Alexander N, Cartwright[2], (1)Dept. of Chemical and Biological Engineering, University at Buffalo (SUNY), Buffalo, NY, (2)Dept. of Electrical Engineering, University at Buffalo (SUNY), Buffalo, NY

Advances in microfluidics and lab-on-a-chip technology have accelerated the development of compact and portable biosensors for point-of-care applications. A relatively new and promising approach to such sensors involves optofluidics where optic and fluidic functionality are integrated at the microscale to leverage their combined advantages. Microfluidic functionality enables compact and rapid processing of small specimens, and optical functionality enables high detection sensitivity of target biomaterials within these samples. In this presentation we introduce a novel all-optical microfluidic biosensor concept. The sensor consists of substrate with an array of embedded microchannels carrying fluid flow perpendicular to its surface. There is a central microchannel and secondary microchannels symmetrically aligned with a periodic spacing about this channel. The central microchannel is illuminated with a focused beam of light propagating perpendicular to the substrate. The periodic spacing of the channels and contrast in refractive index between the fluid and the substrate act to confine and guide the incident light down through this channel. The transmission spectrum through the sensor exhibits minima that depend on the dimensions, periodicity and refractive index of both the carrier fluid and the substrate material. Sensing is achieved via accumulation of a thin layer of target biomaterial on the surface of the microchannel walls, which are functionalized to bind with the material as it flows through the system. The presence of the biolayer causes a shift in the optical transmission minima. We study this effect at optical wavelengths as a function of device parameters and material properties. We also use computational fluid dynamics to model fluid flow through the sensor. Our analysis demonstrates that detectible shifts in the transmission spectrum can be achieved with nanoscale accumulation of biomaterial within the sensor and with relatively little refractive contrast between this material and carrier fluid. The transmission mode operation of the sensor can enable multiplexed biosensing on a single sensor platform.

See more of this Session: Biosensor Devices II
See more of this Group/Topical: Topical 9: Sensors

(288g) Speaking Truth to Power

Deborah Grubbe, BP Products North America, Inc., Chadds Ford, PA

Speaking truth to power

See more of this Session: Developing Leadership Skills
See more of this Group/Topical: Management Division

(235b) Plasmon-Enhanced Nanoscale Metamaterial Sensors

D. Keith Roper, Chemical Engineering, University of Arkansas, Fayetteville, AR, Phillip Blake, Department of Chemical Engineering, University of Arkansas, Fayetteville, AR, Gyoung-Gug Jang, Ralph E. Martin Department of Chemical Engineering, University of Arkansas, Fayetteville, AR and Braden Harbin, Microelectronics-Photonics, University of Arkansas, Fayetteville, AR

Plasmon-enhanced Nanoscale MetamaterialSensors

Presenting Author: D. Keith Roper. University of Arkansas,dkroper@uark.edu

Nano Bio Photonics Lab, Chemical Engineering, http://comp.uark.edu/~dkroper/

Microelectronics/Photonics Graduate Program, http://microep.uark.edu/

Coauthors: Phillip Blake; Gyoung Jang; Braden Harbin

Nanoscalemetamaterials exhibit coupling between plasmon-active elements that createsstrong, local electromagnetic fields. These fields enhance sensitivity morethan 10-fold relative to localized surface plasmon (nanoparticle) or plasmon polariton(BIACore) sensors. **Plasmons** are oscillating free electrons confined tometal surfaces excited by incident electromagnetism at specific wavelengths (1).But sensitivity gains from plasmon field enhancements are in *limitedregions* of the electromagnetic spectrum and only occur in *closeproximity* to plasmon-active nanostructures. These limitations arise from thewavelength-specificity and nanometer-scale localization of plasmon resonances,respectively. **Metamaterials**, whose tunable electromagnetic functionalityderives from three-dimensional (3D) structuring of suitable condensed-mattercomposites, can provide *delocalization* of plasmon effects as well astunability across the electromagnetic spectrum to permit far-field, broad-spectrum,plasmon enhancement. However, progress in developing plasmon-active nano-scalemetamaterials for sensing is limited by (1) lack of comprehensive, multi-scale *models*to evaluate plasmon-exciton coupling and design ordered plasmon-active, 3Dmetamaterial nanostructures; (2) limited availability of inexpensive, bottom-upmethods to fabricate plasmon-active metamaterials for sensing applications; and(3) scant analytic and experimental evaluation of plasmon-active nanoscale metamaterialssensing. We present the first sensor data from nanoscale metamaterialsfabricated using inexpensive, scale-able methods guided by solutions of NavierStokes[1]and Maxwell-Stefan equations. We characterize sensor properties using TEM,SEM, T-UV, and Raman spectroscopy. Data corresponds to theoretical sensor resultsbased on analytic and finite difference time domain (FDTD) solutions to Maxwell's[2] equations. A new,refined metric is introduced to compare sensitivity of nanoscale metamaterialswith nanoparticle and BIAcore sensors on an equivalent basis.

[1] Jang, G.-G. and D.K.Roper. Continuous flow electroless plating enhances optical features of Aufilms and nanoparticles. *J. Phys. Chem C.* (2009) 113:19228-19236.

[2] Roper, D.K., Ahn,W., Taylor, B., D'Asen, Y. Enhanced spectral sensing by electromagneticcoupling with localized surface plasmons on subwavelength structures. *IEEESensors* (2010) 10(3) 531-540.

See more of this Session: Micro and Nanofabricated Sensors I
See more of this Group/Topical: Topical 9: Sensors

(286c) Computational Modeling and On-Sun Model Validation for a Multiple Tube Solar Reactor with Reflective Cavity Walls

Janna Martinek[1], Carl Bingham[2] and Alan W. Weimer[1], (1)Department of Chemical and Biological Engineering, University of Colorado, Boulder, CO, (2)National Renewable Energy Laboratory, Golden, CO

A three-dimensional steady state computational fluid dynamics model is developed for a solar receiver consisting of a reflective cavity with a windowed aperture enclosing an array of five tubes. The computational model couples radiative transfer with fluid flow, heat transfer, mass transfer, and chemical reaction kinetics. The accuracy of the finite volume (FV) method for the approximate solution of the radiative transfer equation is examined for a 2D cross-section of the receiver for the solar and emitted energy components on each of four unstructured grids with three sets of control angles and constant pixelization. FV model accuracy was judged by comparison with Monte Carlo solutions. FV results for the solar component either failed to approach the MC solution or required highly refined spatial and angular grids to achieve accurate, non-oscillatory solutions whereas solutions for the emitted component showed only a weak dependence on the angular and spatial grids.

Based on these results, radiation heat transfer is included in the full 3D model via a combination of ray tracing, Monte Carlo, and finite volume methods. Ray trace modeling of the concentrating system provides the magnitude and direction of the solar energy incident on the receiver window surface. Transport of solar radiation in the cavity is decoupled from all other transport processes occurring in the receiver and profiles of the absorbed solar radiation are determined via a Monte Carlo model using only the receiver geometry, magnitude/direction of radiation incident on the window, and spectral/directional optical properties. A finite volume model is implemented in conjunction with the overall CFD model to account for thermal radiation emitted by heated surfaces in the receiver. Absorption coefficients, scattering coefficients, and the scattering phase function for the cloud of particles in each tube are determined via Mie theory. Transport of the aerosol particles is modeled using a single fluid mixture model with a population balance including transport by convection, Brownian motion, and thermophoretic diffusion. Gasification of 40nm carbon particles is used as a test reaction with Arrhenius kinetic parameters from the literature. Temperature distributions and, correspondingly, reaction conversions among the five tubes are found to be considerably non-uniform, with large temperature gradients developing between the front and back surfaces of each tube. Validation of the base heat transfer and reaction models is accomplished using experimental data taken on-sun at the High Flux Solar Furnace (HFSF) at the National Renewable Energy Laboratory (NREL) with both inert and reacting materials reaching temperatures up to 1400°C.

See more of this Session: Concentrated Solar for Power Generation and Chemical Processing I

(11g) Glucose Biosensor Based On Encapsulated Enzyme with Peptide Nanotubes

Byung-Wook Park[1], Kyoung-A Ko[2], Do-Young Yoon[3] and Dong-Shik Kim[1], (1)Department of Chemical and Environmental Engineering, University of Toledo, Toledo, OH, (2)Department of Pharmacy, University of Toledo, Toledo, OH, (3)Chemical Engineering, Kwangwoon University, Seoul, South Korea

Application of bio-inspired nanostructures to enzymes can enhance the performance of enzymes and enzymatically facilitated systems. Peptide nanotubes (PNTs) are used to maintain enzyme activity that is susceptible to the enzyme orientation and surrounding conditions. The enzymes encapsulated inside a PNT may have higher thermal stability, long-term storage stability and resistance to denaturants compared to the free enzymes in a solution. PNT's nanostructure has the environmental functionalities such as biocompatibility, large surface-to-volume ratio and robust backbone structure. Moreover, the unique properties of PNTs provide a possible prospect such as electrical conductivity. In this study, a nanostructure of the peptide nanotube was examined for possible use as a support material for enzyme encapsulation. A model enzyme, hydrogen peroxidase (HRP), was efficiently bound into the peptide nanotube. The catalytic activity of HRP encapsulated inside PNTs was examined in terms of thermal stability, long-term stability, and resistance to a denaturant. The results showed that the encapsulated HRP has a higher enzyme activity than the free HRP in a solution. Moreover, an amperometric biosensor was developed using a bi-enzyme system of glucose oxidase (GOx) and the encapsulated HRP into the PNTs to detect glucose. The results showed great potential of the encapsulation technique in the development of more stable biosensors without compromising the enzyme sensitivity. These results suggest that the encapsulation of enzyme inside a PNT can be used as a stabilization technique in biocatalyst and biosensor applications where retaining the enzyme activity is important.

See more of this Session: Biosensor Devices I
See more of this Group/Topical: Topical 9: Sensors

(512a) Academics and Open Innovation

Richard L. Long, New Mexico State University, Las Cruces, NM

Open Innovation generally refers to property rights issues associated with orgainzations such as nine sigma. In a broader sense, it refers to the issue of "credit" of which cash awards are one type. The issue of "control" and the financial and other rewards have to do with the relationship between buyer and seller. Transactions must have a degree of confidence in the integrity of the parties and establishing that confidence is a key requirement.

See more of this Session: Challenges In Open Innovation and Growth
See more of this Group/Topical: Management Division

(235c) Novel Optical Sensors Utilizing Surface Plasmon Enhanced Transmission Via Diffraction Gratings with Controlled Nanostructured Topology to Probe Non-Absorbing and Absorbing Adsorbates

Wei-Hsun Yeh[1], Andrew C. Hillier[2] and Joseph W. Petefish[1], (1)Department of Chemical and Biological Engineering, Iowa State University, Ames, IA, (2)Iowa State University, Ames, IA

The interaction of light with nanostructures, particularly those containing noble metals, can produce a variety of unique and useful optical phenomena. Optical nanostructured sensors based on SPR have several advantages in comparison to those based on fluorescence-labeled agents in detection assays. Examples of benefits from SPR-based sensors include high sensitivity, exclusion of expensive and complex labeling processes, and in-situ real time monitoring of adsorption processes at the interface. Enhanced transmission through nanostructured gratings is one method SPR can be applied for sensing.

Grating-based optical sensors utilizing surface plasmon enhanced transmission have gained increased research interest in comparison to traditional prism-coupled configurations. The collinearly simple yet precise setup of spectrophotometer systems for grating-based sensors have superior potential for developing portable devices; traditional prism-coupled configurations require more complicated alignments. In this work, we demonstrate the enhanced transmission SPR via well-defined and controlled nanostructures of diffraction gratings. Not only were observed lager enhanced transmissions, but various matching conditions of SPR can be fine tuned by simple rotation of samples with respect to incident p-polarized light.

We also developed a novel gold-coated chirped grating in which various topologies are incorporated into a single diffraction grating. Notably, this grating demonstrates the impact of various local topologies, i.e. combinations of pitch and amplitude, on the responses of enhanced transmissions. In addition, we demonstrate the capabilities of this novel chirped grating as a substrate for high-performance optical sensing.

Since many adsorbates absorb light, a direct method to probe the existence and thickness of absorbing films utilizing SPR is necessary. Although SPR-based sensors mainly probe the thickness of non-absorbing adsorbates, there is interest in exploring the effect of molecular resonances (MR) in absorbing adsorbates on the SPR response. We also demonstrate the ability of SPR-based grating sensors to probe the interaction between MR and SPR. Notably, an enhanced transmission peak near the proximity of the Q band of a chromophore was observed. In other words, this enhanced transmission peak implies "transparent metal induced by MR and SPR". In summary, this work demonstrates the utility and versatility of grating-based SPR sensing techniques.

See more of this Session: Micro and Nanofabricated Sensors I
See more of this Group/Topical: Topical 9: Sensors

(11f) Protein Array for Activity Screen of EGF Receptor (EGFR) Tyrosine Kinase Inhibitors In Cancer Cells

Gargi Ghosh and Sean P. Palecek, Department of Chemical and Biological Engineering, University of Wisconsin-Madison, Madison, WI

Protein tyrosine kinases are critical for regulating important cellular processes. Deregulation of kinase activity resulting from receptor overexpression, gene amplification, and increased enzymatic activity can lead to several diseases including cancer. As a result, protein kinases have emerged as promising drug target and several inhibitors are at different stages of development. Hence, there is an urgent need of a simple, robust yet cost-effective platform for high-throughput screening of small molecule drug candidates. A major challenge in developing kinase assay involves sensitive and specific detection of enzyme activity in the background of multiple tyrosine kinases present in cellular extracts and obtaining accurate measurements from small biopsy samples.

Here we report the development of protein array for quantitative profiling of epidermal growth factor receptor (EGFR) kinase activity. EGFR signaling pathways have been implicated in playing an important role in cancer development and progression. Many therapeutics are at different stages of development which are directed towards EGFR kinase domain. The extent of substrate phosphorylation upon incubation of EGFR substrate with lysates of cancer cells corresponded to the difference in activation status of EGFR in different cancer cell lines. Additionally, the ability of the diagnostic tool to quantify the extent of inhibition of EGFR activity by different tyrosine kinase inhibitors establishes the potential of this tool to screen for the most promising therapeutics for individual patients and monitor treatment progression.

See more of this Session: Biosensor Devices I
See more of this Group/Topical: Topical 9: Sensors

(11a) DNA Hybridization Detection Using Liquid Crystals

Stephanie M. Malone and Daniel K. Schwartz, Chemical and Biological Engineering, University of Colorado at Boulder, Boulder, CO

Sensitivity to hybridization is the basis for genetic screening techniques such as fluorescence in-situ hybridization (FISH), as well as microarray based assays. Elimination of the fluorescent marker, scanner, and the requisite labor necessary to label the DNA target, however, could allow for the fabrication of cheap and disposable point-of use devices. Previous work demonstrated that double stranded DNA aligns 5CB and MBBA liquid crystals (LC) at 30 degrees to the extension direction, while the LC aligns parallel to extended single stranded DNA. This difference in alignment is sufficient to detect a single DNA molecule. Detection of double stranded DNA over the background of single stranded DNA will determine the sensitivity of this approach. By orthogonally labeling double and single stranded DNA it was demonstrated that LC alignment is sensitive enough to detect hybridized DNA among single stranded probe molecules. In situ detection of hybridization was also explored.

See more of this Session: Biosensor Devices I
See more of this Group/Topical: Topical 9: Sensors

(308a) Micromachined Multielectrode Microprobes for Neurotransmitters with An On-Probe Iridium Oxide Reference

Vanessa M. Tolosa[1], Tina T.-C. Tseng[1], Kate M. Wassum[2], Nigel T. Maidment[2] and **Harold G. Monbouquette**[1], (1)Chemical and Biomolecular Engineering, University of California, Los Angeles, Los Angeles, CA, (2)Psychiatry and Biobehavioral Sciences, University of California, Los Angeles, Los Angeles, CA

We are designing, producing, and implementing micromachined, multielectrode microprobes for the monitoring of neurotransmitter concentration changes in near real-time in the brains of live rodents. Our devices may be used to investigate a variety of neurological disorders *in vivo*. For example, glutamate and dopamine are neurotransmitters whose imbalance has been linked to Huntington's disease and Parkinson's disease as well as addiction. We construct 150 microprobe devices simultaneously on four-inch silicon wafers with four, micron-sized electrodes per probe using micro-electro-mechanical-systems (MEMS) fabrication technologies. One or more microelectrode sites per probe is chemically modified with permselective polymer films and immobilized glutamate oxidase for the electroenzymatic detection of glutamate with sensitivity of 120 nA $\times$ uM^{-1} $\times$ cm^{-2}, detection limit of 2 uM, and 1 sec response time. A second site may be modified with a thin layer of electrodeposited overoxidized polypyrrole for sensing of dopamine (DA) with sensitivity of 3250 nA $\times$ uM^{-1} $\times$ cm^{-2}, detection limit of 60 nM and 1 sec response time. Both sensors are highly selective against the array of electrooxidizable species normally present in brain extracellular fluid. The electrodeposition of an iridium oxide (IrOx) film on a third site provides an on-probe reference. Such an on-probe reference eliminates the need for a separate reference electrode, which traditionally has been a macroscopic Ag/AgCl wire electrode that causes inflammation of living tissue. We have observed the additional benefit of a ~60% decrease in signal noise *in vivo* using the on-probe IrOx reference. Although the IrOx reference exhibited pH-dependence of potential corresponding to -77 mV/pH, the extracellular pH in brain tissue normally is maintained tightly over the range of ~7.35 to 7.45. Thus, our microprobes with IrOx reference were able to detect successfully a physiologically relevant range of glutamate concentrations in the presence of interferents (*e.g.*, ascorbic acid, dopamine) both *in vitro* and *in vivo*.

See more of this Session: Micro and Nanofabricated Sensors II
See more of this Group/Topical: Topical 9: Sensors

(254b) Solar Thermochemical Production of Solar-Grade Silicon: Thermodynamic and Economic Analyses

Ronald Michalsky[1], Peter Pfromm[1], Bryon Parman[2] and Vincent Amanor-Boadu[2], (1)Department of Chemical Engineering, Kansas State University, Manhattan, KS, (2)Department of Agricultural Economics, Kansas State University, Manhattan, KS

Solar thermochemical production of solar-grade silicon:thermodynamic and economic analyses

RonaldMichalsky [1], Peter H. Pfromm [1], Bryon Parman[2], Vincent Amanor-Boadu [2]

[1] Department of Chemical Engineering, [2]Department of Agricultural Economics,

Kansas StateUniversity, Manhattan, Kansas, USA

Availability oflow cost solar-grade silicon will improve the sustainability of electricitygeneration from photovoltaic thin-film-Si cells. A precursor for manufacturingsolar-grade Si, polycrystalline Si is conventionally produced via carbothermalreduction of silica above 1700 C in an electric arc furnace with carbonelectrodes. Besides the use of electricity as a costly energy source, theprocess emits fossil CO_2 at least equimolar to the amount of Siproduced. Sustainability would be greatly improved if the release of fossilcarbon as CO_2 indirectly due to electricity generation and directlyfrom silicon reduction could be reduced or eliminated.

Solar-thermal H_2production via H_2O hydrolysis oxidizing a Zn reactant and sequentialZnO thermal dissociation at up to 2000 C using concentrated solar radiation isa well studied solar thermochemical approach. Materials such as Al, $Si_3N_4$and SiC have been produced from their metal ores via solar carbothermalreduction.

To reduce theenergy expenses during Si production and to avoid associated CO_2emissions, this work proposes solar thermochemical reduction of SiO_2at high temperatures using sulfur as reducing agent. Production of elementalsulfur is outpacing demand in the marketplace raising issues with sulfurdisposal. Based on Gibbs free energy computations, a process sequence forproducing polycrystalline Si and sulfuric acid from SiO_2, elementalsulfur, H_2, air and water at near ambient pressure and hightemperatures is proposed here. Technical considerations such as temperature requirementsand the avoidance of the cumbersome handling and mixing of two solids will bediscussed. To assess sustainability and economic competitiveness the mass andenergy balance for the conceptual process will be presented along with an economicanalysis determining the price of sulfur-based solar-grade Si required forbreaking even with plant investment and operational costs.

See more of this Session: Process Development and Design for Solar-Grade Silicon Production
See more of this Group/Topical: International Congress on Energy 2011

(286e) Techno-Economical Analysis of Solar Thermochemical Ammonia Production At near Atmospheric Pressure

Ronald Michalsky[1], Peter Pfromm[1], Bryon Parman[2] and Vincent Amanor-Boadu[2], (1)Department of Chemical Engineering, Kansas State University, Manhattan, KS, (2)Department of Agricultural Economics, Kansas State University, Manhattan, KS

Techno-economical analysis of solar thermochemicalammonia production at near atmospheric pressure

RonaldMichalsky [1], Peter H. Pfromm [1], Bryon Parman[2], Vincent Amanor-Boadu [2]

[1] Department of Chemical Engineering, [2]Department of Agricultural Economics,

KansasState University, Manhattan, Kansas, USA

Ammonia is thebasis for modern agriculture and ensures the food supply for a growing worldpopulation. It is used widely as base chemical for the chemical industry. Over100 million metric tons NH_3 are produced world wide per year. NH_3has been proposed recently as sustainable transportation fuel and may enableefficient H_2 storage.

At industrialscale the Haber-Bosch process synthesizes NH_3 catalytically at high pressure and elevated temperature with naturalgas and nitrogen as main inputs. Globally this process consumes up to 5%of all natural gas produced and 2% of the total world energy production, withsignificant fossil-based CO_2 emissions. Aiming at reactive NH_3synthesis without natural gas from air and steam at ambient pressure, this workstudies a solar thermochemical cycle employing a molybdenum-based reactant.

Based on Gibbsfree energy computations, a process sequence of Mo_2N formation fromits elements, nitride hydrolysis forming NH_3 and MoO_2 atambient pressure, and metal oxide reduction using solar heat and H_2as chemical reducing agent below 1200 C, is proposed. Experimental data on aMo-based transition element reactant utilized for the fixation of $N_2$and formation of NH_3 from the solid nitride hydrolyzed with steamwill point out process limiting steps such as diffusion in the solid state andionicity of an interstitial nitride reactant. To assess sustainability andeconomic competitiveness compared to the state of the art Haber-Boschprocess, the mass and energy balance for Mo-based solar thermochemical NH_3synthesis including on-site production of H_2 used as reducing agentwill be presented along with an economic analysis determining the price of NH_3required for breaking even with plant investment and operational costs.

See more of this Session: Concentrated Solar for Power Generation and Chemical Processing I
See more of this Group/Topical: International Congress on Energy 2011

(34d) Patenting Basics

Vivek Koppikar, Patent Attorney, Dale City, VA

This partially invited session will provide a fundamental review of the patent procurement process and patent law in general, and is designed for anyone who has any type of interest in the patenting process and/or deals with patents in their work in industry and/or academia. Little or no background knowledge in patents is needed by those intending to attend this session. The session will begin with the very basics of the patenting process as well as providing them with basic terminology. Next, the session will provide a basic synopsis of the patent procurement process with the US Patent and Trademark Office. The session will then focus on portions of a patent most relevant in patent disputes and patent litigation. Case studies involving various patent scenarios will be presented. The session will wrap up with a question and answer session where attendants can ask basic questions dealing with the patent procurement process and patent law in general. This session will provide a foundation for two following patent sessions focusing first on drivers for innovation in energy technology and, next, on litigation for engineers. Abstracts are solicited for case studies from lawyers and engineers with various patent experience

See more of this Session: Law, Risk, and Ethics of Technology and Project Management
See more of this Group/Topical: Management Division

(254c) The Horizontal Ribbon Growth of Solar Silicon Crystals: Process Analysis, Stability, and Control

Parthiv Daggolu[1], Andrew Yeckel[1], Carl Bleil[2] and Jeffrey J. Derby[1], (1)Chemical Engineering and Materials Science, University of Minnesota, Minneapolis, MN, (2)Ribbon Technology International

Present-day photovoltaic technology is predominantly based on the fabrication of solar cells from silicon. The cost and efficiency of silicon-based solar cells depend primarily on the growth rate and degree of single-crystallinity attained by the growth process. Traditional silicon crystal growth methods are either very expensive or compromise on efficiency. The horizontal ribbon growth (HRG) process is a promising crystal growth technology that has the potential to overcome these limitations. William Shockley first conceived of the HRG process in late 1950's, with subsequent efforts by Kudo in Japan in the late 1970's and by Energy Materials Corporation in the US in the early 1980's. After encouraging initial development, these efforts stalled owing to various technical difficulties.

We are applying a comprehensive thermal-capillary model to study the coupled phenomena of heat transfer and interfacial phenomena (solidification and capillarity) in the HRG process. This model accounts for heat transfer in the melt-crystal-crucible domains with radiative heat loss from high temperature surfaces, melt convection due to buoyancy and surface-tension forces, and the self-consistent determination of melt-crystal, melt-ambient and crystal-ambient interface shapes. This moving-boundary problem is solved numerically by the Galerkin finite element method, using elliptic mesh generation.

Extensive parametric sensitivity studies are conducted to identify the variables that impact the process. The effect of pull rate, pulling angle, melt height, crucible geometry and furnace heat transfer are investigated. Transient simulations are performed to analyze the dynamics of the growth process and to determine its stability. Preliminary results indicate that quasi-steady-states exist for this system but that most are unstable. In this light, the dynamic analysis of this system will be require to identify attainable stable states and possible strategies to stabilize unstable ones.

See more of this Session: Process Development and Design for Solar-Grade Silicon Production
See more of this Group/Topical: International Congress on Energy 2011

(641f) Secondary Thermochemical Xylo-Oligomer Hydrolysis of High-Solids Dilute-Acid Pretreatment Slurries

Erik M. Kuhn[1], Joseph Shekiro[1], Nick J. Nagle[2] and Richard T. Elander[1], (1)National Renewable Energy Laboratory, Golden, CO, (2)National Bioenergy Center, National Renewable Energy Lab, Golden, CO

Dilute-acid pretreatment is a critical step in the biochemical conversion of biomass to fuels and chemicals, releasing hemicellulose as oligomeric and monomeric xylose and rendering the cellulose fraction readily digestible by enzymes. It has been difficult to achieve very high yields of monomeric xylose during pretreatment, since acid hydrolysis also converts xylan to various oligomers and non-fermentable sugar degradation products. An alternative strategy for the production of high yields of monomeric xylose is to perform secondary thermochemical hydrolysis on the dilute-acid pretreated material. This secondary hydrolysis, typically performed at lower temperatures than the primary dilute-acid pretreatment step, liberates additional monomeric xylose while minimizing the yield of sugar degradation products. This work presents the results of laboratory and pilot scale primary and secondary thermochemical hydrolysis experiments over a wide range of conditions. Our results suggest that reduced severity conditions for primary dilute-acid pretreatment provide the highest overall yields of monomeric xylose after secondary thermochemical hydrolysis for the conditions tested. The results of our LC-MS investigations suggest that higher-molecular weight xylo-oligomers are converted to xylose during secondary thermochemical hydrolysis, but that a small fraction of the xylan in corn stover is converted to stable but non-fermentable compounds, such as xylo-oligomers, during both primary and secondary hydrolysis.

See more of this Session: Chemical and Catalytic Conversions and Processes for Renewable Feedstocks
See more of this Group/Topical: Sustainable Engineering Forum

(288f) Effective Communications

Gautam Pillay, Department of Chemical Engineering, Rowan University, Glassboro, NJ

Effective communication between management, employees, external stakeholders, and partner organizations (often geographically separated).

See more of this Session: Developing Leadership Skills
See more of this Group/Topical: Management Division

(288h) Dhamma Tools for Managers

Gavin Towler, UOP, Des Plaines, IL

This paper will describe methods for maintaining your composure, motivating your team, dealing with obstacles, getting results and becoming a better leader, while maintaining a harmonious working environment.

See more of this Session: Developing Leadership Skills
See more of this Group/Topical: Management Division

(288e) Dealing with Uncertainty In Decision Making

Diana Matonis, Flucomp Enterprises, Hickory Hills, IL

What is the difference between risk and uncertainty?

What are some quantitative methods that aid decision making under uncertainty?

What are some qualitative methods that aid decision making under uncertainty?

How can decision makers effectively combine qualitative and quantitative methods?

See more of this Session: Developing Leadership Skills
See more of this Group/Topical: Management Division

(512c) Exogenous Uncertainty: Choosing Innovation Projects

Diana Matonis, Flucomp Enterprises, Hickory Hills, IL

What is the difference between risk and uncertainty?

What are some quantitative methods that aid decision making under uncertainty?

What are some qualitative methods that aid decision making under uncertainty?

How can decision makers effectively combine qualitative and quantitative methods?

See more of this Session: Challenges In Open Innovation and Growth
See more of this Group/Topical: Management Division

(10a) Fermentation Cost Improvement Through Propagation Advances for a Commercially Relevant Biomass-to-Ethanol Process

Kate Brandon Sutton, B. Emme, L. Jones, S. McHatton, C. Kang, C. Corbett, L. Putnam and M. Torry-Smith, Novozymes North America Inc., Franklinton, NC

Compared to their starch-to-ethanol counterparts, pentose-utilizing organisms for biomass-to-ethanol processes are less robust while having to ferment mixed sugars in a significantly more inhibitory substrate. This inhibition can generally be overcome with a high yeast pitch to maximize xylose uptake rates in fermentation. The cost driven trade-off of a high yeast pitch during fermentation is the necessity of dedicated onsite propagation, as opposed to purchase of all yeast. Novozymes' Advance Biofuel cost model suggest the majority of the fermentation cost when using high yeast pitches can be attributed to the propagation variable costs, most significantly from the carbohydrate source. Therefore, maximizing the cost performance of C5 yeast propagation is important to ensure an economically feasible biomass-to-ethanol fermentation step.

Novozymes has identified two approaches to reduce the fermentation cost: first, increase propagation cell yields from sugars to reduce the realized organism cost, preferably on cheap carbohydrate sources, or second, produce a more robust organism from propagation with a higher realized xylose uptake rate in fermentation which would enable fermentation at a lower yeast pitch in a reasonable processing time. This presentation will discuss cost model results in conjunction with reactor data to pinpoint potential savings and opportunities to lower overall fermentation cost for second generation ethanol. Sensitivities on cost from propagation and fermentation process variables targets will be presented. Propagation optimization with respect to higher realized xylose uptake rate in fermentation will be addressed as well. Lastly, the holistic minimum ethanol selling price implications for process optimizations will be given.

See more of this Session: Biological Conversions and Processes for Renewable Feedstocks I
See more of this Group/Topical: Sustainable Engineering Forum

R&D Management from Beginning to End

Mukund Karanjikar
Senior Associate
Technology Holding LLC
Salt Lake City, Utah

Jack Hipple
Principal
Innovation-TRIZ
Tampa, FL

Eldon Larsen
Professor of Engineering
Marshall University
South Charleston, WV

Extended Abstract

This workshop focuses upon managing R&D projects from beginning to end. Beginning refers to mental picture of an idea whereas end refers to solving the customers need and in the process generating profit for the organization. Using a case study, the speakers will delineate challenges involved at every stage of R&D project from a mental image of an idea to a profit making product. The talk also compares old paradigm of innovation management with the new proposed paradigm. The probability of success can be significantly improved using the new paradigm.

Successful Innovation Management

We have seen many idea management flow sheets, gate management processes, and prioritization recommendations starting with an idea that already exists, but no discussion of where the idea came from or what was necessary to generate and frame it. Moreover attention is not paid to shaping and propagation of the idea. This has been especially true over the past decade when processes such as Six Sigma came into vogue whose main thrust was to eliminate deviation. New concepts and innovation are, by their very nature, deviant! In order to generate fresh and innovative ideas we must create a new thought pattern that may challenge the existing status quo in any number of ways. These can even include, at an extreme, replacing one's current products and businesses with totally new approaches and models.

The key elements of successful commercial innovation include:

1. Continuous flow of unmet needs, both from current customers and related to core competencies of the organization ("outside in" as well as "inside out"
2. Using clear mental models of problems and potential solutions
3. Perseverance for idea propagation
4. Clear definition of the business opportunity
5. An appropriate product development and management team
6. Efficient process scale up and continued input from the market
7. Successful project management to insure implementation on schedule and within budget

We will look at each one and discuss the important issues.

1. Continuous Flow of Unmet Needs and Exploration of Uses for Core Competencies

An organization needs to have a process for how it decides what to work on. This may sound simplistic, when this is left to intuition or accident; it is unlikely that the "best" projects will be chosen.

Steps to make sure that an innovation effort is capturing the best opportunities:

- Talk to groups that are not your current customers but that might be able to use the functional capability that you have as a core competence. Customer competitors are another valuable source. The customers of your customers are another excellent source. This insures that you have awareness several levels above your product and customers and alerts you to potential long term needs and trends that may not be obvious to your direct customer or may not be readily disclosed by them.

Figure 1: Level Diagram

In these discussions we want to make sure that we have a clear definition of what the "ideal result" that our customers are desiring. We can also look at parallel universes of technology and markets to explore opportunities.

2. Using Clear Mental Models of Problems and Solutions

It is one thing to "invent" a new product, but another to make sure that we understand how a potential customer is going to use the product as well as how the customer's customer is going to use their product. The days are gone when we can just make something, put it in a drum or tank car, and wait for the invoice to be paid. We must be intimately familiar with the product's use, its limitations (including regulatory issues if appropriate). We must understand the details of the physical use of the product and how it affects a customer's products and business.

It is also important to make sure that we do not have blinders on when thinking about use of our product. It is rare that there is only one way to accomplish a function. We must be constantly vigilant about competitive activities, not just from direct competitors, but from parallel approaches to meeting a customer's needs.

3. Perseverance for Idea Propagation

New business and product development is hard work and it is easy to get discouraged. An idea that starts our with a 10% probability of success can raise its odds to over 75% by the proper, early stage identification of what challenges must be overcome. If these are not clearly defined, then a new product or business development will flounder and struggle. The perseverance battle rests more on the shoulders of individuals than the organization or top management. Idea has to propagate and the best person to do so is the idea owner. Ideas may be facilitated top-down (by management) however propagation is always bottom up (by employees). A proprietary tool (YouSNAP®) is being proposed to better manage idea propagation. The five elements of the tool are:

a. You – Know yourself
b. Surroundings – Know your surroundings
c. Network – Develop your network
d. Approach – Decide an approach
e. Persistence – Be persistent and iterate

Employees need to understand a simple truth. Ideas are dime a dozen and never bought on face value. Frustration is part of innovation. One should know that failures will exist. Hence know yourself. Are you up to it? Then you should know your surroundings. Often times, a tendency exists in corporations to blame the corporation itself. What is corporation? It is made up of people like yourself trying to serve the customers. If there are non-cooperative people around you, the journey would be commensurately difficult. However the important thing is that you should know this before. This leads to having a robust network of supporters. Once you have done the necessary preparation, decide upon an approach. Finally be persistent. The presentation will discuss the tool in more interactive format.

4. The Business Opportunity

A new product is not a business unless we have a way of sustaining the product's sales, expanding its utility, and developing and enforcing an intellectual property portfolio. As we commercialize a new product, we must be constantly aware of, and be willing to address with resources and commitment, issues such as change in product form, packaging and shipping issues, product purity, and shelf life/stability. These issues are most adequately addressed by constant and continuous communication with the customer.

5. Product Development Teams

We frequently choose product team members based on "who's available". While this may be expedient, it may not serve the best interests of the new business development. The obvious component of team selection is function—marketing R&D, manufacturing, legal, sales, etc. but we frequently ignore the human side of team functioning. If team members spend too much time dealing with personal issues relating to other team members, that energy is not applied to successful new product launch. Tools that can assist in this include measurements such as the Myers Briggs family of instruments, the Hermann Brain Dominance instrument (HBDI), and the Kirton KAI assessment. Knowledge of these personality profiles will not change natural behavior, but can make team members aware of others' preferred styles and open discussion can eliminate hidden barriers to effective team functioning.

6. Efficient process scale up and continued input from the market

Often times, companies strive for launching the best product. It is important to realize that continued input from the market is critical in order to have that 'best product' in the market. It is necessary to efficiently scale up and launch a product sooner than later. It should be perfected using market feedback. Getting market feedback is the only way to close the R&D loop and serve your customers to the fullest.

7. Successful Project Implementation and Management

New product development does not end when the first semi-commercial sample has been OK'd by the customer. A new type of team is now required to oversee the details process design, plant construction (including all the sub issues of permitting, EPA regulations, etc.). Quick and efficient plant start up with appropriate contingency planning is also a part of this effort. Contingency planning, not only related to unforeseen plant construction and start up difficulties, but also to taking care of the customer's needs in these situations must be part of the plan.

Three keys to successful project management lead to successful project completion. These keys involve

- The importance of people
- The importance of defining and setting good objectives
- The importance of good planning

We will apply these three keys to successful project management. This is all about application, not just theory. Today, you can apply the lessons you will learn.

A case study will be discussed in the presentation.

(491b) Progress In Biomass Trait Screening Using a High Throughput Digestibility Platform

Nick Santoro[1], Shane Cantu[2] and Jonathan Walton[2], (1)Great Lakes Bioenergy Research Center, Michigan State University, East Lansing, MI, (2)Michigan State University, East Lansing, MI

A key step in the development of a viable lignocellulosic ethanol industry is the improvement of the plants that are to be used as biomass feedstocks. We have previously described a high throughput cell wall digestibility platform (HTDP) to facilitate the identification and evaluation of feedstock plants. Cell wall digestibility is assayed by the amount of free Glc and Xyl released after chemical pretreatment and incubation with a commercial cellulase mixture (Accellerase). One key component of the HTDP is a custom-designed workstation that grinds and dispenses individual plant biomass to within 0.1 mg accuracy. Combining this workstation with a second key component, a laboratory information management system (LIMS), provides the basis for generating a quantitative measure of cell wall digestibility on a vast number of samples.

The results presented in this talk summarize cell wall digestibility screening projects carried out in the two years since the HTDP opened. A collection of approximately 7,000 *Brachypodium* EMS mutant M2 lines have been screened. Classes of outliers are identified and studies are underway to further characterize the mutant lines. A collection of approximately 1,200 *Arabidopsis* T-DNA lines with insertions in cell wall biosynthesis-related genes have been screened for differences in cell wall digestibility. The chemical and genetic characterization of these lines is presented. Corn lines comprising a University of Wisconsin Diversity Set have been screened recently using corn stalk core samples. An initial survey of outliers is presented. Results for transgenic poplars containing knock-down insertions for decreasing the expression of the CCR gene are also presented. The poplar screening studies compare the cell wall digestibility of the transgenic and wild type poplar in response to a variety of pretreatment conditions as well as to geographical field location. Substantial differences in cell wall digestibility of the transgenic and wild type poplar are shown as the above parameters are varied.

In Summary, the HTDP is an extremely useful tool, demonstrating utility as a screening tool to identify improved plants from large sample populations (Brachypodium). In addition, the HTDP is flexible to accommodate novel sampling methods (corn stalk cores). Furthermore, once interesting plants are identified, they can be further tested in the HTDP using a number of varied pretreatment conditions (poplar).

See more of this Session: Advances In Biofuels: DOE Bioenergy Research Centers I
See more of this Group/Topical: Sustainable Engineering Forum

(288c) Strategies for Motivating Employees

Charlene A. Wall-Warren, BASF, Florham Park, NM

Strategies for motivating employees

See more of this Session: Developing Leadership Skills
See more of this Group/Topical: Management Division

(61a) The Future for 3M in Nanomaterials Science and Engineering

Larry A. Wendling, Corporate Research Laboratory, 3M, St. Paul, MN

3M's fundamental business model can be described as technology-driven innovation. Having 45 technology platforms, 3M ranks among industry's R&D leaders in terms of its breadth of technical capabilities. The ability to access novel size dependent properties via nanoscale materials science and process engineering potentially impacts the majority of all our technology platforms and provides significant opportunities for differentiated product performance across many of 3M's businesses. Recent advances in nanomaterials science and engineering at 3M providing novel optical, mechanical, electrical, and catalytic properties will be described.

See more of this Session: 2011 Annual Meeting Plenary - Chemical Science Innovation: Future of the US Chemical Enterprise I (co-sponsored by the Council for Chemical Research)
See more of this Group/Topical: Liaison Functions

(61b) Educating Global Chemical Engineers in the 21st Century

Robert A. Brown, Boston University, Boston, MA

Chemical engineering thrived in the United States during the 20th century because of the confluence of discoveries in chemistry, innovations in process technology, the abundance of raw materials, and markets fueled by economic growth. What will chemical engineering in the United States look like in 2050? Will we experience the same level of innovation and prosperity in an economy that is driven by substantial growth off our shores? This presentation will discuss the impact on the chemical engineering discipline of the emergence of China, India and other countries as economic equals. Some potential strategies will be offered for professional and institutional leadership in the globalized economy that lies ahead.

See more of this Session: 2011 Annual Meeting Plenary - Chemical Science Innovation: Future of the US Chemical Enterprise I (co-sponsored by the Council for Chemical Research)
See more of this Group/Topical: Liaison Functions

Corning's R&D / Innovation Approach

October 17, 2011
American Institute of Chemical Engineers

Joe Miller, Ph.D.
Executive Vice President
Chief Technology Officer

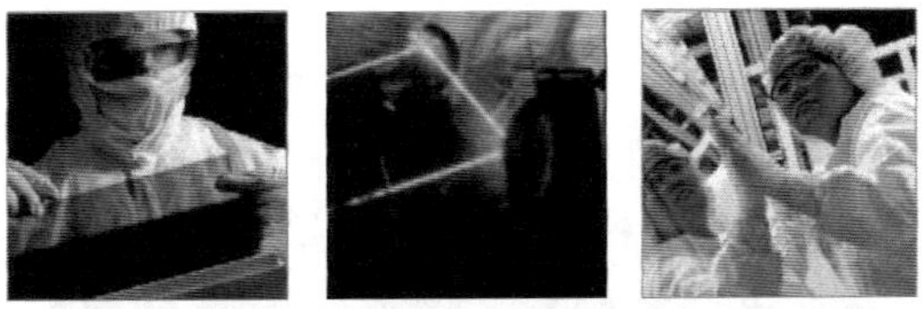

Corporate Snapshot

CORNING

Corning Incorporated
Who We Are

- Founded: 1851
- R&D Lab Established: 1908
- Headquarters: Corning, NY
- Employees: ~26,000 globally
- Sales: ~$6.6 billion
- Sales (incl. pro rata JV sales): ~$11 billion
- RD&E @ 9% of sales $670 million

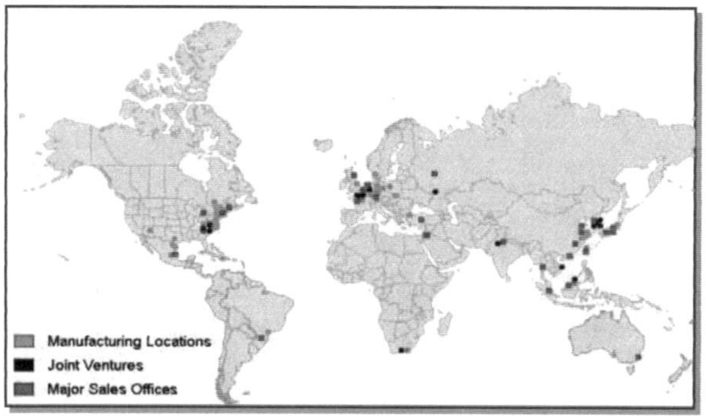

- Corning's global operations:

  ~ 65% of revenues from outside North America

  More than 50% of our employees are outside the United States

Corning Incorporated
Who We Are

Market Segments

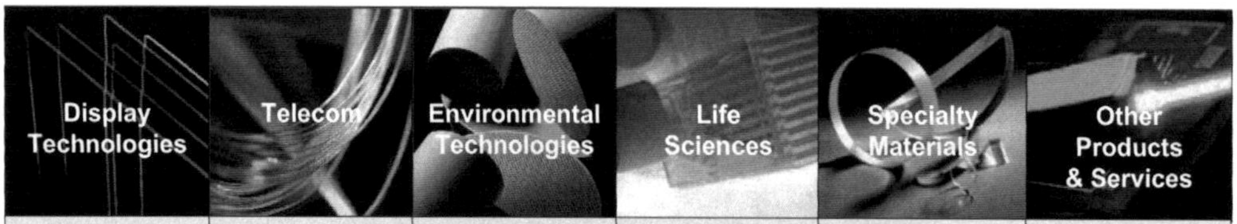

Display Technologies	Telecom	Environmental Technologies	Life Sciences	Specialty Materials	Other Products & Services
• LCD Glass Substrates • LTPS-LCD Glass Substrates	• Optical Fiber & Cable • Hardware & Equipment	• Emissions Control Products – Automotive – Diesel – Stationary	• Cell Culture & Bioprocess • General Laboratory Products	• Advanced Optics & Materials • Display Optics & Components • Semiconductor Optics & Components • Aerospace • Astronomy • Optical Metrology • Ophthalmic • Telecom Components • Specialty Glass	• Display Futures • New Business Development • Drug Discovery Technology • Equity Companies – Dow Corning Corp. – Samsung Corning Precision Glass Company, Ltd – Eurokera, S.N.C

Grow from the Confidence of 160 Years of Innovation

Who We Are

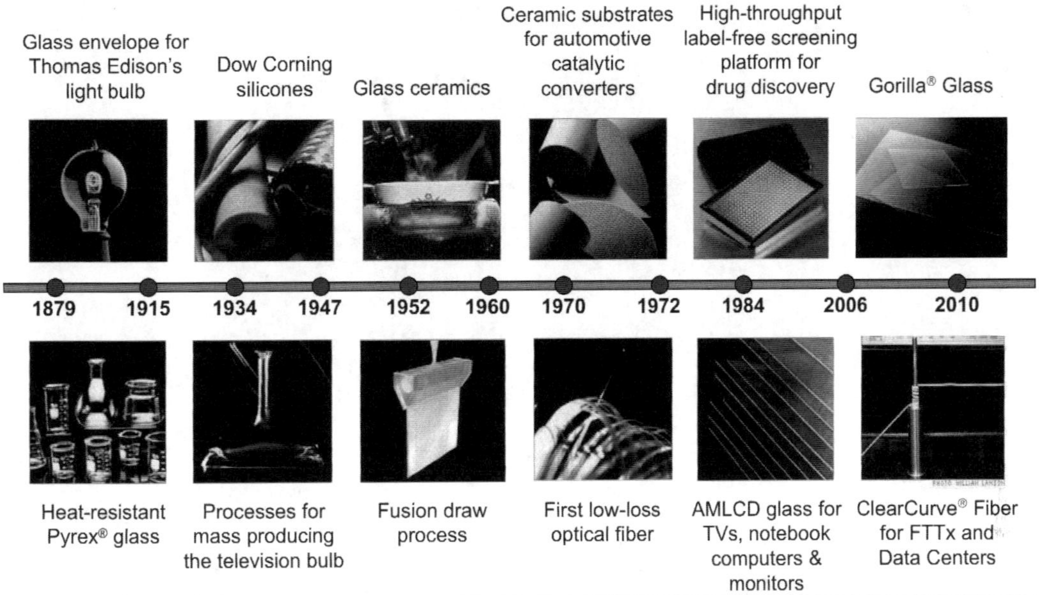

CORNING

5

Economic Disruptions Evolution of Strategy

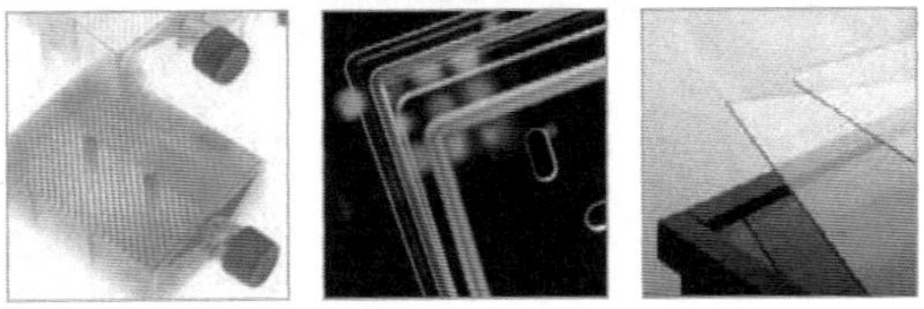

CORNING

6

Look where we started 8 years ago
Where have we come from?

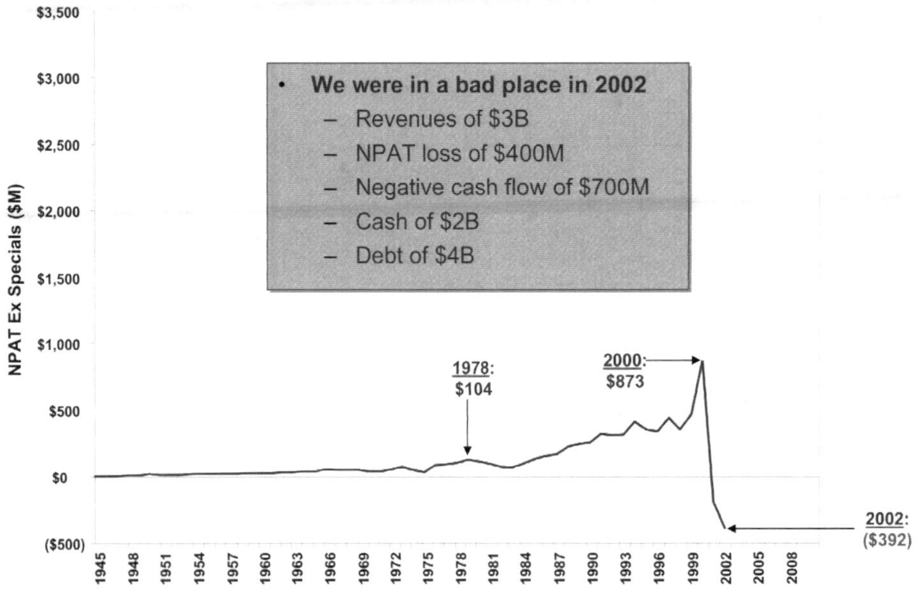

CORNING © Corning Incorporated 2008 | 7

Corporate Strategy Framework
Another 150 Years of Innovation & Independence

We Need To Grow	***We Also Need Stability***
Grow Through Global Innovation	**Provide Stability and Balance**

Synergy ⇄ Tension

We Must Preserve Trust

Live the Values

CORNING © Corning Incorporated 2008 | 8

Restructured and Aligned RD&E with Strategy Framework

- Eliminated worldwide network of satellite labs
- Concentrated R&D in Corning, NY, and Fontainebleau, France
- Reduced total RD&E spending by 50%
- Focused on strengthening core capabilities
  - Glass
  - Ceramics
  - Optical Physics
  - Process Engineering
- Added new competencies carefully – ones that added uniqueness when combined with our materials and process core
- Concentrated on supporting the growth of ongoing businesses
- Created an early stage business group to prime the innovation pipeline
- Added new innovation governance system

... and then began to reinvest in growth through innovation

CORNING

Where have we come from?

We've delivered outstanding financial performance

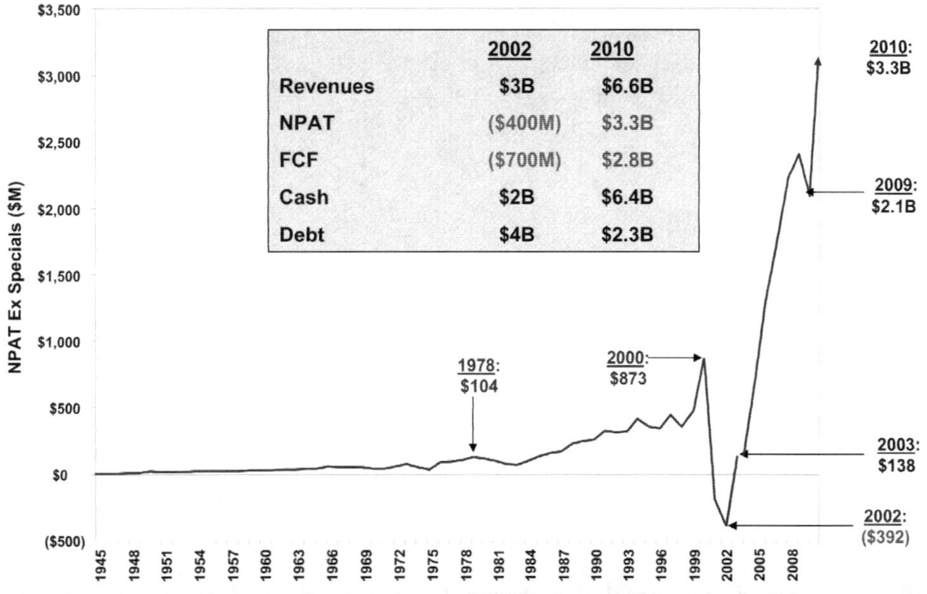

	2002	2010
Revenues	$3B	$6.6B
NPAT	($400M)	$3.3B
FCF	($700M)	$2.8B
Cash	$2B	$6.4B
Debt	$4B	$2.3B

CORNING

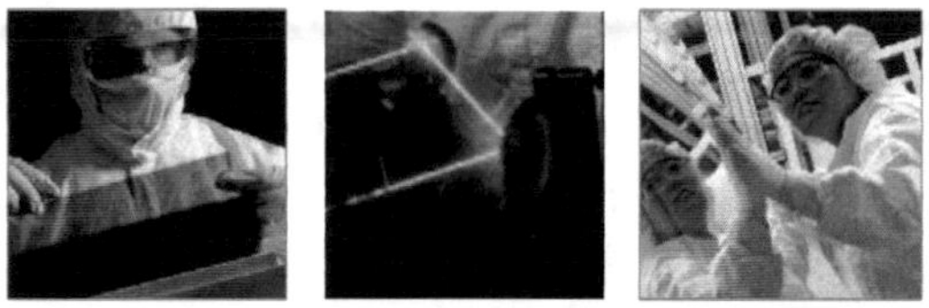

Strategy Implementation
The Innovation Recipe

CORNING

Corning Has A Distinctive Innovation Recipe

- Take on tough problems
- Stick to our competencies
- Achieve dominant design
- Build strong IP barriers
- Capacitize quickly
- Sell to concentrated markets with high seller power
- Multiple JV's for access
- Significant RD&E and capital investments

Created one to two businesses/decade

CORNING

Innovation Recipe Yields Keystone Components

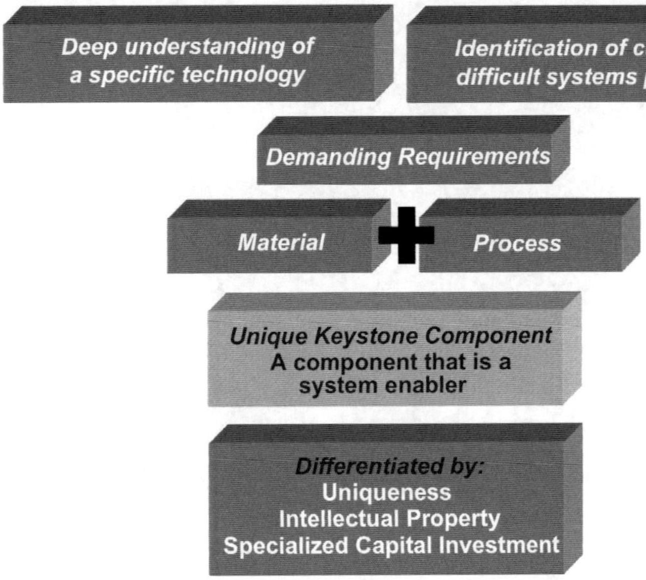

CORNING © Corning Incorporated 2008 | 13

Keystone Components
Based on Integration of Multiple Competencies

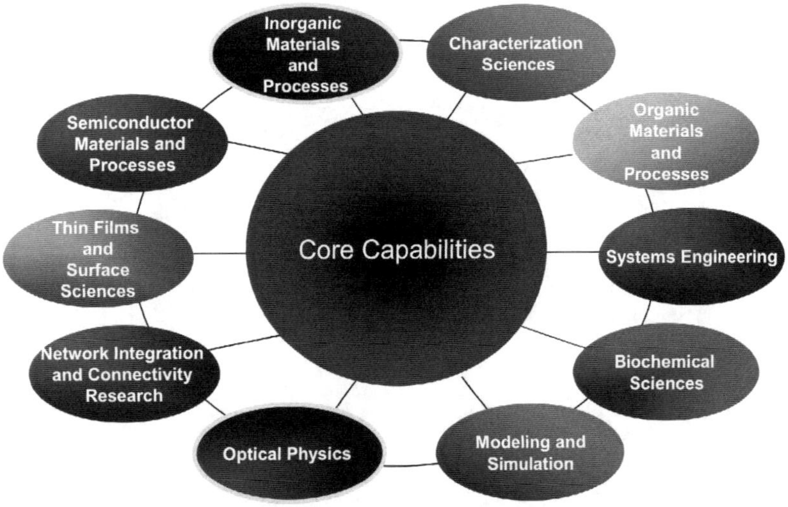

CORNING © Corning Incorporated 2008 | 14

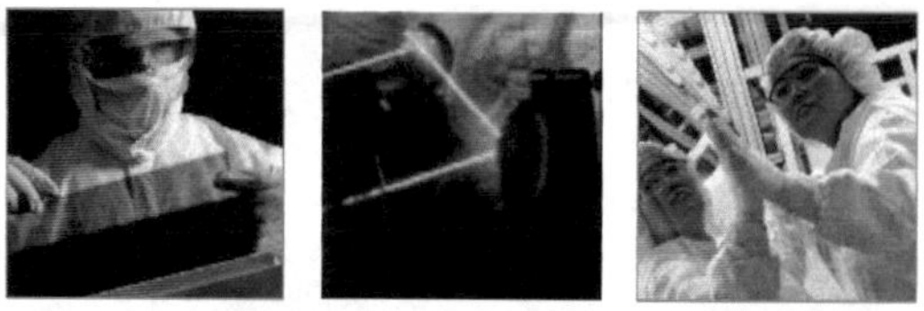

Innovation Strategy
Response to 2001

CORNING 15

We consciously centralized our Research & Development activities and facilities

CORNING 16

Central R&D Facility at Sullivan Park – Corning, NY

Sullivan Park Connected Globally in Europe, Asia and the U.S. West Coast

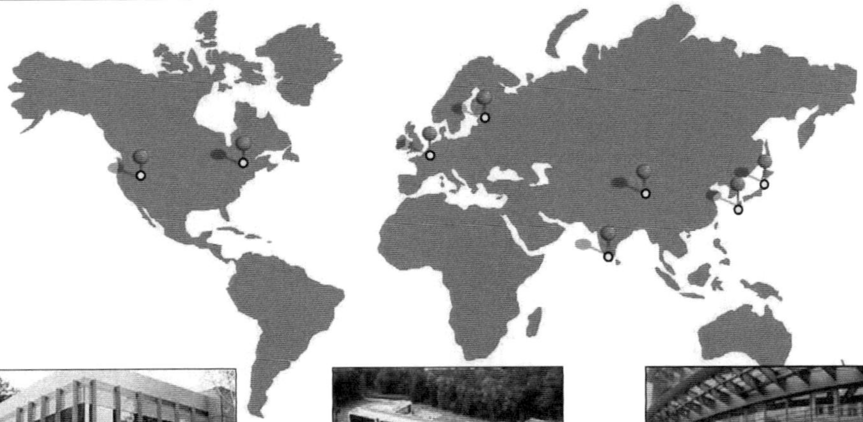

Corning West Technology Center
Silicon Valley, California

Corning European Technology Center
Fontainebleau, France

Corning Advanced Technology Center
Taipei, Taiwan

Hardened our Core

Strategic Approach to Competencies

Dig Deeper in Historic Competencies

- Precision melting, fusion, extrusion, and draw
- Glass, glass ceramics, and cellular ceramics
- Optical physics

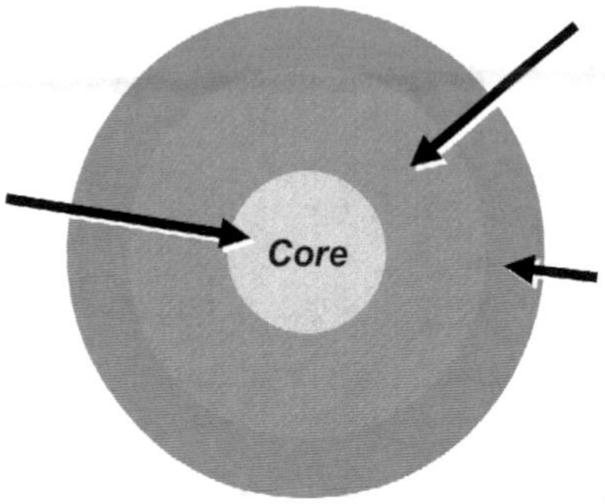

Add New Competencies

- Optical Semiconductors
- Biophotonics
- Carbon

Enabling Capabilities

- Process and materials modeling
- Organics
- Characterization
- Finishing
- Coating
- Systems engineering
- Assay development

Created New Corporate Innovation Governance

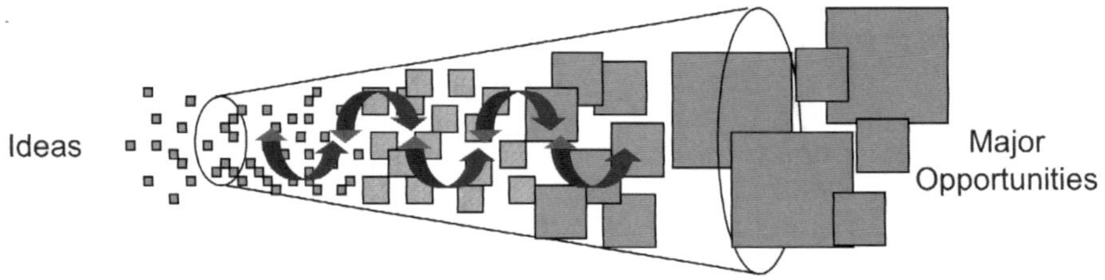

Corporate Technology Council

- **Create-the-Next Wave Portfolio**
- **Early stage opportunities**
- **CTO + Research & NBD Leaders**

Growth Execution Council

- **Invest-in-the-Future Portfolio**
- **Sort, pace and execute**
- **CEO, CTO + CTO Staff**

Technology & Strategy with the Board

Reinforced Five Stage Innovation Process

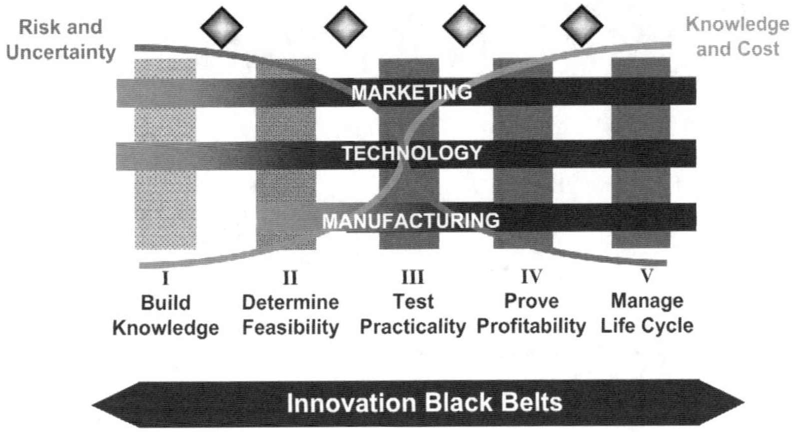

Research & Development Today

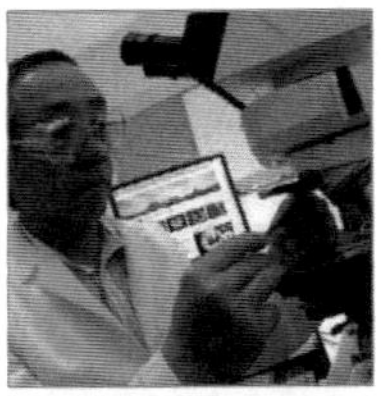

- RD&E investment intensity ~ 9% of sales
  - 5% Development & Engineering
  - 4% Research & New Business Development
- Core Competence R&D Model
- Centralized R&D Laboratory Model
  - Sullivan Park R&D Center, Corning, New York
  - 1900 employees (technical + support)
  - 40 different countries represented
- Global reach-out in Europe and Asia

Effective Technology Delivery to Current Businesses

Display	Diesel	Auto	Telecom	Specialty	Life Scie...e
• LCD Substrates • Poly-Silicon Substrates • Thin Substrates	• Light Duty Diesel Filters • Heavy Duty Diesel • Stati...	• Next Gen Substrates 600/2 400/3 • Thin Wa...	• FTTx – Fiber, Cable, H&E • ClearC... ...ltra Low - loss Fiber • Pretium® EDGE - Data Center	• Gorilla™ ... ...s for KrF • Sensor Products • Precision Optics	...stem Assays • Advanced Cell Growth Surfaces

~Two New Products Each Year for Each Business

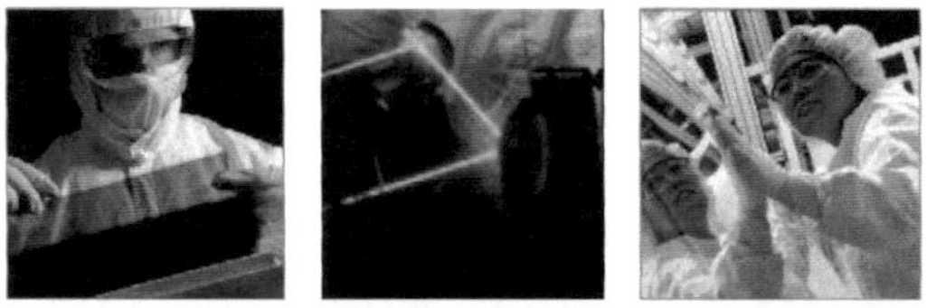

Innovation Strategy Response to 2009

Adjacencies - The Third Strategic Leg

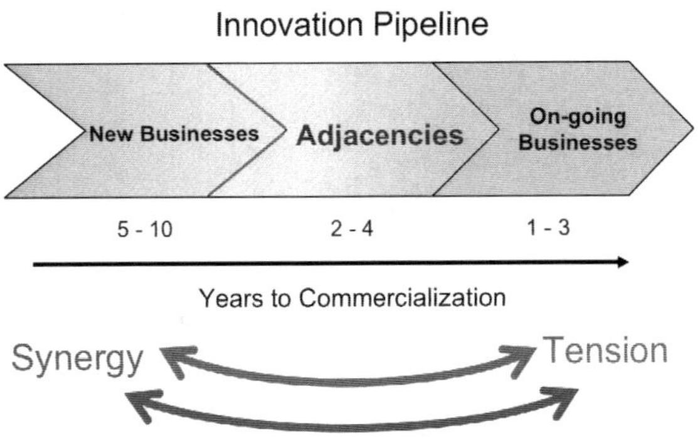

Forced the new leg into the middle

With adjacencies, we tuned our strategy to create more balance in the portfolio using existing technology assets

- Specialty glass is our core materials and process capability
- We have substantial glass sheet forming process technology and manufacturing assets to support the Display Technologies business
- We redeployed some of those assets to address adjacent market segments in Consumer Electronics with thin, strong, damage-resistant glasses
- Here's an illustration of the adjacency concept

Illustrative

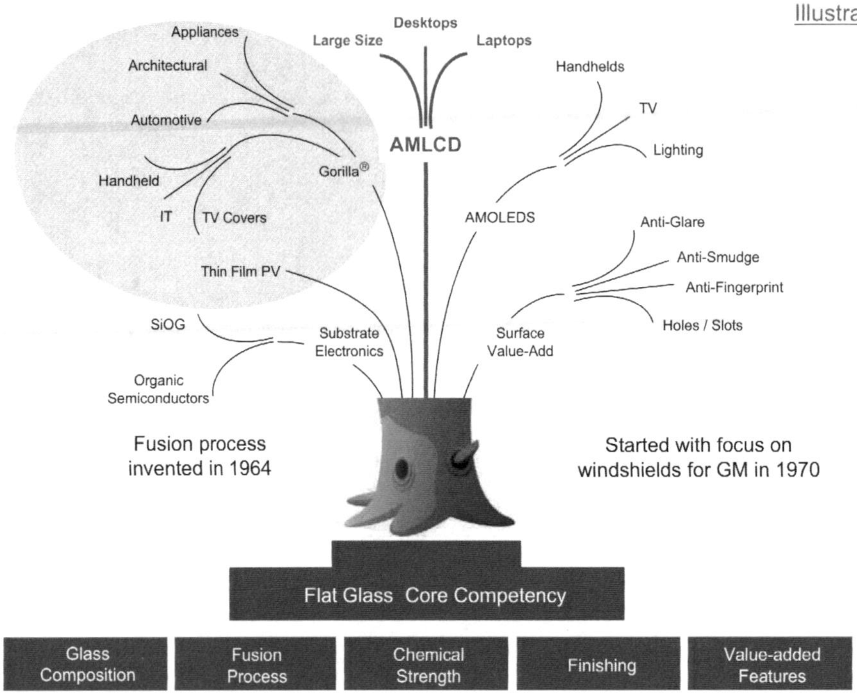

2011 Strategic Frame

- Corporate technology strategy has three legs
  - Support the ongoing businesses near term
  - Build an adjacency opportunities portfolio medium term
  - Maintain exploratory activities and new business development longer term

- Make room for significant shift to adjacencies
  - Resource the fastest growing segment – glass
  - Restructure New Business Development
    - Prune the New Business Development project portfolio
    - Direct some of those resources to adjacencies
  - Maintain exploratory research to create a pipeline of opportunities

2011 Innovation Portfolio Investment Allocation

Near Term	Medium Term	Long Term
Existing Businesses • Display: Gen 10 • Diesel: AT/AC/CO • Telecom: FTTx; Next ... **60%** • Touch Substrates FIT (Full-sheet Integrated Touch) • Thin Film PV	Advanced Life Sciences Advanced Consumer Electronics **25%** Advanced WIN/Wire... Optical Communications Energy & Environmental	Reset in Stage I • Mercury Abatement New Opportunities **15%** • Native Green Lasers • Advanced Batteries • CO_2 Sorption
1-3 years	2-4 years	5-10 years

2011 Innovation Portfolio

Near Term (60%)	Medium Term (25%)	Long Term (15%)
Major Businesses • Display • Diesel • FTTx & Next-Gen Data Center **Flat Glass Adjacencies** • Gorilla® Glass - Handhelds/IT Covers - Touch Screens - TV Covers • Thin-Film Photovoltaics	Advanced Life Sciences Advanced Consumer Electronics and Industrial Applications Advanced Wire/Wireless Optical Communications Energy & Environmental	• Organic Semiconductors • Thermoelectrics • Advanced Batteries • CO_2 Sorption • Ceramic Adjacencies • Optical Adjacencies
1-3 years	2-4 years	5-10 years

Gorilla® Glass Applications
Touch Devices

- "Projected capacitive" touch devices are changing the game
  - Multi-touch
  - More complex applications and gestures
  - Better optical performance
- Touch device growth accelerating
  - Handheld 20% CAGR
  - Notebook/netbook 60% CAGR
- Potential for sensor integration on cover glass

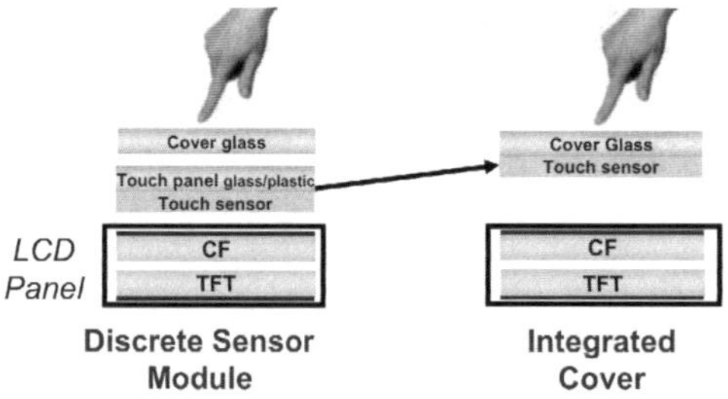

Gorilla® Glass - The Slate Effect

- Glass will play a significant role in this space
  - Thinner LCD glass
  - Thinner Touch sensor glass
  - Thinner, yet stronger cover glass will be required
- Weight and area increase 5 to 7x on slates vs. handheld devices
- System designers faced with significant challenges
- Gorilla a perfect fit for reliability

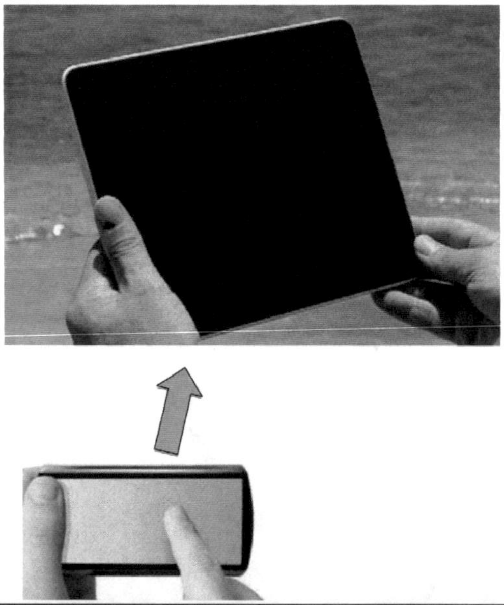

Glass Adjacency Innovation Results

- Gorilla® Glass is exceeding growth expectations
- Growing awareness glass as a new design element
- More than 30 major brands using Gorilla® Glass
- Next generation technology delivering significant improvements in both performance, capabilities and product possibilities
- With a potential to achieve close to $ 1B revenues in 2011

CORNING

Strategic Response Delivered Strong Technology Performance

- Technology leader in each of our market segments
- More than 50% of revenues over last three years from new products
- Corning leads the industrial materials segment in intellectual property performance
  - > 200 patents granted per year
  - 400 – 500 patent applications in each of the past two years

CORNING

Innovation Wisdom
What I've Learned

Dynamics of Innovation – Observations

- Innovation is more than a process
- Innovation is an interconnected system
  - Linked to strategy
  - *Supported* by process
  - Directed by leaders
- Talented people enabled to do their best - a critical success factor
- Innovation must go beyond today and look to New Markets / Creating New Businesses

Principles / Philosophy – My Top 10

- Strive for balance
- Pursue multiple avenues to maximize value from R&D investment
- Know "who you are" and what your "recipe" is
- Select early stage projects based on broad criteria together with a large dose of experience
- Mistakes happen – don't try to eliminate them

CORNING | Science & Technology Corning Restricted | 37

Principles / Philosophy – My Top 10

- Great leaps are very risky
- Effective change is evolutionary …requiring continuous evaluation and feedback processes
- Business silos achieve focus and intensity - their long term preservation leads to a steady decline in technology enrichment, breadth and connection
- R&D must be aligned with business strategy but have enough independence to balance today's needs with the opportunities of the future
- Don't forget the social aspects – people have to work well together

CORNING | Science & Technology Corning Restricted | 38

Other Bits of Innovation Wisdom - I

- The number of ideas for a robust pipeline >>> number than most believe are necessary
- Idea generation must be global with multiple sources and connections
- An unstaffed bench of ideas and opportunities promotes greater realism in assessment of current projects
- Multi-disciplinary teams of competent, motivated people beat individual genius every time … more productive, moderate risk, and compress cycles
- Sufficient rate of new business creation requires both linear and non-linear innovation pipeline flow
- Technology, marketing, manufacturing in it from the start

You get the business where you got the business

Other Bits of Innovation Wisdom - II

- Balance the deep scientific competencies with deep engineering competencies – both are needed for success
  - Great outcomes from meshing the engineering mindset with the scientific mindset
- Inject advanced engineering into the early stages of the innovation process to create and adapt more effective process invention
  - Process invention has been equally important as material invention for Corning's innovations
  - Corning embedded the advanced engineering group in the R&D center and hired PhD engineers in multiple engineering disciplines
- Reduce the learning cycles on process engineering with pilot scale laboratory operations
  - Goal: Every major manufacturing process platform in the R&D lab

Other Bits of Innovation Wisdom - III

- Modeling and simulation … and materials and process characterization are critical enablers for materials and process engineering success
  - Yield better fundamental understanding at all levels of the innovation process, create strong IP positions, and often lead to significant cost reduction
- Faster clockspeed innovation opportunities require more technology presence at the customer interface
  - Reduce the learning cycles and requirements iterations
  - Application development, applications engineering and product/process engineering are prominent roles at the customer interface
- Gross margin (GM) performance is really the name of the game
  - GM% is what actually funds RD&E activities
  - Process engineering innovation plays a major role in improving GM% and extending profitable business life cycles

Talent … Talent … Talent

- RD&E people are different than the prevailing business culture
- Continuous infusion of new talent is refreshes any organization, especially R&D
- Development of well balanced innovation leadership demands multiple experiences as early as possible
- Teams prevail
- Different talent is needed at different stages
- T-shaped people with depth and breadth

Successful Innovation
The Critical Role of Leadership

- Capabilities, processes, culture, and money are critical

- But it takes *individual will* to lead innovation
  - **determination and resolve in the face of uncertainty and change**
  - **acceptance of failure and comfort with scientific risk**

- *And individual skill*
  - **the backgrounds of our top leaders must reflect**
    - **real skills in technology and science**
    - **understanding – and rich intuition -- of how and why technology creates business value**

CORNING | Science & Technology Corning Restricted | 43

At the End of the Day...........

- Innovation is Easy to say, Hard to do
  - It is all about the "how": culture, leadership, talent, and process … and patience … and money

- A balanced innovation investment
  - Grows the existing businesses
  - Creates new businesses

- Still plenty of opportunity for invention and innovation in glass, ceramics and other inorganic materials
  - Especially when integrated with other technologies

CORNING | Science & Technology Corning Restricted | 44

CORNING

(61c) Novel Structures for Public-Private Partnerships: Case Studies with Medical Devices and Biopharmaceuticals

Sangtae Kim, Morgridge Institute for Research, Madison, WI

At a time when the growth rate in research opportunities towers over the funding capacity of traditional sources of sponsorship, universities are turning to translational research and the associated licensing revenue streams as the untapped resource to close the gap. This hope is challenged by the historical track record which shows that most universities do not have a successful track record in technology transfer. In this context, this presentation examines the current landscape for technology transfer via spin-off companies generated from university research. The classical view of a "basic research phase" and a "VC-supported development & commercialization phase" separated by the so called "valley of death" albeit simplistic, offers some insights into the high failure rates. The presentation proposes a new model for a public-private partnership with the goal of higher success rates for translational research and development. The model will be illustrated with recent case studies involving medical devices and biopharmaceuticals.

See more of this Session: 2011 Annual Meeting Plenary - Chemical Science Innovation: Future of the US Chemical Enterprise I (co-sponsored by the Council for Chemical Research)
See more of this Group/Topical: Liaison Functions

(137a) The Future of Fuels and Alternative Feedstocks - Recognizing Hype vs. Practical Limitations

William Banholzer, Dow Chemical Company, Midland, MI

The world now agrees that the ideal future requires 100% sustainable sources of energy and product feedstocks in adequate amounts to support a high standard of living for all.

Which new pathways and technologies will emerge to transform our situation? This question is addressed from the perspective of the chemical industry, which was built on oil, natural gas, and coal. These have served as the major raw material feedstocks and energy sources for driving reactions and separations. The industry is now shaping its transformation to sustainable energy and is developing new materials and solutions for energy supply and conversion.

Here we consider the mass and energy balances, capital investment and resource requirements of several key alternative energy and feedstock technologies. These considerations determine where we can expect realistic progress toward sustainable chemistry in both the short and long term, and where we should place our investments.

See more of this Session: Annual Meeting Plenary - Chemical Science Innovation: Future of the US Chemical Enterprise II (co-sponsored by the Council for Chemical Research)
See more of this Group/Topical: Liaison Functions

(137b) Importance of Human Capital: Why Major in Chemical Engineering?

John Anderson, Illinois Institute of Technology, Chicago, IL

Innovation is driven by people who are smart, creative and assertive. One strategy for innovation in the chemical enterprise is to attract people like this to chemical engineering and chemical sciences. Opportunities for careers in technical fields are much different for today's students than five decades ago when I chose a major. Compounding the challenge is that engineering in general has lost market share of students at universities. This presentation attempts to stimulate thinking about steps that higher education and industry can take to attract future innovators to the chemical enterprise and meet their educational objectives.

See more of this Session: Annual Meeting Plenary - Chemical Science Innovation: Future of the US Chemical Enterprise II (co-sponsored by the Council for Chemical Research)
See more of this Group/Topical: Liaison Functions

(137c) The Future for Emerging Energy Technologies

Anthony Cugini, U.S. Department of Energy, National Energy Technology Laboratory, Pittsburgh, PA

The Department of Energy's Office of Fossil Energy, through the National Energy Technology Laboratory, supports a variety of R&D initiatives to develop alternative clean fuels and chemicals from fossil sources, including coal, natural gas and biomass. A key R&D element of the current Fuels Program is the development and demonstration of advanced separation membranes for the production of hydrogen in advanced IGCC power systems for subsequent use as a turbine fuel or conversion to environmentally benign fuels and chemicals. Other research areas focus on the development of modifications necessary for combined coal/biomass conversion which include co-feeding to the gasifier, evaluation of the differences in reaction kinetics and reaction products of the gasification process and the effects of additional biomass contaminants on downstream water gas shift (WGS) and Fischer-Tropsch (FT) conversion catalysts. This presentation will provide an overview of DOE's ongoing R&D initiatives for the co-production of power and fuels/chemicals from coal and coal/biomass mixtures in advanced IGCC systems.

See more of this Session: Annual Meeting Plenary - Chemical Science Innovation: Future of the US Chemical Enterprise II (co-sponsored by the Council for Chemical Research)
See more of this Group/Topical: Liaison Functions

(137d) Chemical Product Life Cycle Management and Enterprise Environmental Stewardship

Monty Alger, Air Products and Chemicals, Inc., Allentown, PA

This presentation will analyze how sustainability affects the chemical industry. The industry has moved from compliance with safety, environmental, and quality requirements to proactive stewardship of products throughout their life cycles. Sustainability -- the integration of environmental, social, and economic considerations -- has become an important element of business management and product development strategy.

See more of this Session: Annual Meeting Plenary - Chemical Science Innovation: Future of the US Chemical Enterprise II (co-sponsored by the Council for Chemical Research)
See more of this Group/Topical: Liaison Functions

(60a) The Importance of Basic Research Investment to Economic Growth

Thomas W. Peterson, National Science Foundation, Arlington, VA

The National Science Foundation is the primary source of support for basic research and education in science and engineering throughout the US academic community. At NSF, the Directorate for Engineering has historically occupied a unique and interesting space within the Foundation, and today is no different. Like other directorates, most of ENG investments support basic research and discovery through its core programs. But a portion of the ENG portfolio of investments directly addresses the important *translation* of the fruits of successful basic research into products and processes of societal benefit. What can one federal agency (the NSF) reasonably do to stimulate innovation and economic development through strategic investments in our nation's colleges and universities?

As we celebrate the International Year of Chemistry, we use this opportunity to examine the societal impact of research investments made by NSF, focusing especially on chemistry and chemical engineering.

See more of this Session: The Year of International Chemistry Plenary
See more of this Group/Topical: Liaison Functions

(272a) F = m a of Biology?

Michael W. Deem, Bioengineering, Rice University, Houston, TX

I will describe my participation over the past few years in an effort to find fundamental mathematical laws in biology. I will discuss a theory for understanding the emergence of multi-scale, hierarchical structure in biology. I will describe how modular structure in proteins, genetics, and biological networks may arise and how it can be understood. I will show the implications of the theory for engineering design, and I will discuss additional examples of structure formation in ecological food networks, developmental pathways, physiology, and social networks. I will then turn to theories of the immune system and describe a theory of the immune response to vaccines. I will illustrate this theory by application to design of the annual influenza vaccine. I will use this theory to explain limitations in the vaccine for dengue fever and to suggest a transport-inspired amelioration of these limitations. I will conclude with a discussion of the immune system in bacteria.

See more of this Session: Professional Progress Award Lecture
See more of this Group/Topical: Liaison Functions

(487a) Beyond Water: Expanding the Spectrum of Efficient Large Scale Separations

William J. Koros, School of Chemical & Biomolecular Engineering, Georgia Institute of Technology, Atlanta, GA

Innovations in membrane materials, structures, modules and system design have fundamentally changed the water purification industry over the past four decades. Revolutionary reductions by a full order of magnitude in energy consumption per unit of delivered product were achieved by substituting pressure-driven processes in place of traditional thermally-driven separation systems. Scalability has also been achieved, with state-of-the-art membrane plants delivering over 50 million gal/day of pure water. More recently, this revolution has begun to spread beyond water to more difficult-to-separate gaseous species differing by 0.5Å or less between permeated and rejected components. Oxygen-nitrogen separation, hydrogen and natural gas purification and monomer capture and recycle typify these extensions. Besides more demanding size discrimination than needed for water purification, more aggressive feeds must also be treated in these nonaqueous cases. Realistically expanding the separation spectrum, therefore, requires a new generation of nanostructured materials ranging from pure inorganics to metals and carbons, rather than conventional organic polymers favored for aqueous systems. Revolutionary materials alone are not enough, since evolution of module manufacturing platforms and seamless integration of these more efficient devices into existing systems via process modeling must also occur. Besides membranes, other low energy intensity approaches like sorbents will also be considered to illustrate options for orderly transition paths to a healthy lower energy intensity chemical industry based on conventional and alternative feedstocks.

See more of this Session: 63rd Institute Lecture
See more of this Group/Topical: Liaison Functions

Paper # 99

Lessons Learned From and Economic Impacts of the Fukushima, Japan Disaster

Emmanuel Dada, ChemProcess Technologies (CPT), LLC, League City, TX, Thomas Mensah, Georgia Aerospace Corporation, Atlanta, GA, Derrick Rollins, Iowa State University, Ames, IA, L. Antonio Estévez, Department of Chemical Engineering, University of Puerto Rico, Mayaguez, PR, Otis Shelton, Compliance and Operational Assessments Program, Praxair, Danbury, CT and Joycelyn Harrison, Aerospace, Chemical and Material Sciences Directorate, Air Force, Arlington, VA

The March 10, 2011 earthquake and tsunami in Japan had a devastating impact on the people and industry of Japan. That event led to the Fukushima I Nuclear Power Plant disaster at Fukushima Daiichi, which is one of the largest nuclear incidents perhaps, second only to the 1986 Chernobyl disaster. The disaster has brought into forefront the debates and concerns over the operation, maintenance, and location of nuclear plants worldwide. In our current economic climate, the reality of reduced preventive maintenance, fewer operators and engineers, as well as a significant decrease in critical experience have contributed to a greater likelihood of accidents in many operations and plants on this planet. How do companies, under tremendous pressure to reduce costs, maintain a high commitment to safety? Are governments and global *watch dogs*, such as the International Atomic Energy Agency (IAEA), doing their jobs adequately or are they compromising too much under economic and political pressures? What technology and knowledge are we lacking to keep the general public safe? How should engineering education change to address these questions as well as the moral and ethical responsibilities that come with using and advancing technology. This session will deal with these questions and more in a panel format featuring the 2011 MAC Eminent Chemical Engineers Award recipients.

The STEM Education and Its Impacts on Pipeline for Underrepresented Minorities

Emmanuel Dada, ChemProcess Technologies (CPT), LLC, League City, TX, Thomas Mensah, Georgia Aerospace Corporation, Atlanta, GA, Derrick Rollins, Iowa State University, Ames, IA, L. Antonio Estévez, Department of Chemical Engineering, University of Puerto Rico, Mayaguez, PR, Otis Shelton, Compliance and Operational Assessments Program, Praxair, Danbury, CT and Joycelyn Harrison, Aerospace, Chemical and Material Sciences Directorate, Air Force, Arlington, VA

Over $500 million in the United States is expended annually in the United States towards programs dedicated to encourage students to study engineering yet enrollment in engineering has not changed significantly and in fact enrollment of underrepresented minorities are in a decline with few exceptions. The enhancement of STEM/engineering work force in the USA can be immediately achieved by fostering a sustainable education program for underrepresented students now in college and K-12.

Currently there is an endemic lack of interest in math and science in US high schools. In order to turn things around there must be a renewed focus and emphasis in developing ways to get students stimulated and interested in math and science. Effective implementation of the strategies and tactics embodied in the "***America COMPETES Act***" (http://www.house.gov/apps/list/speech/edlabor_dem/RelJuly22MathScience.html) is required to reverse the trend. The US competitiveness in the world is in not in an immediate danger because of the current strong attraction to talent from outside by US institutions. When conditions improve to the extent that people have the opportunity to remain and engage in rewarding careers in their home countries then the US competiveness will be in seriously impacted. There is strong evidence that this is beginning to happen in countries like Singapore, China, and India. USA Today in its September 21, 2009 issue highlighted some of the reasons for the current exodus of skilled workers from USA which include promising career opportunities in their home countries; family ties to reconnect with their families and culture with support system lacking in the USA; and their frustrations, long delays and uncertain immigration status.

(99c) Q&A and Final Remarks

Emmanuel Dada, ChemProcess Technologies (CPT), LLC, League City, TX, Thomas Mensah, Georgia Aerospace Corporation, Atlanta, GA, Derrick K. Rollins Sr., Chemical and Biological Engineering, Iowa State University, Ame, IA, L. Antonio Estvez, Department of Chemical Engineering, University of Puerto Rico, Mayaguez, PR, Otis Shelton, Compliance and Operational Assessments Program, Praxair, Danbury, CT and Joycelyn Harrison, Aerospace, Chemical and Material Sciences Directorate, Air Force, Arlington, VA

There will be a ten minute question and answer session followed by ten minutes of final remarks from all panelist and closing remarks of 5 minutes.

See more of this Session: MAC Chemical Engineering Forum - Invited Speaker Award Recipients
See more of this Group/Topical: Liaison Functions

(99d) Presentation of Awards

Maria K. Burka, National Science Foundation, Arlington, VA

October 17th presentation of 2011 MAC Eminent Engineers Awards at the AIChE's 2011 Annual Meeting (Minneapolis, MN).

See more of this Session: MAC Chemical Engineering Forum - Invited Speaker Award Recipients
See more of this Group/Topical: Liaison Functions

(205a) Rapid Development of Coal Chemical Industry in China Providing a Rewarding Career for a Chemical Engineer

Zhenping Lu, Shenhua Coal to Liquid Company, Shenhua Group, Shenhua, China

Abstract currently unavailable.

See more of this Session: CACS 30th Anniversary Symposium: Global Career Opportunities for Chemical Engineers
See more of this Group/Topical: Management Division

(205b) Global Career Opportunities for Chemical Engineers

Kelly Seibert, UOP LLC, Des Plaines, IL

Abstract currently unavailable

See more of this Session: CACS 30th Anniversary Symposium: Global Career Opportunities for Chemical Engineers
See more of this Group/Topical: Management Division

(205d) Career Opportunities for Chemical Engineers in India

K. Venkataramanan, Larsen & Toubro Limited, Mumbai, India

abstract currently unavailable

See more of this Session: CACS 30th Anniversary Symposium: Global Career Opportunities for Chemical Engineers
See more of this Group/Topical: Management Division

(205c) Career Opportunities for Chemical Engineers in Singapore

Jim Yang Lee, Chemical Engineering, National University of Singapore, Singapore, Singapore

Abstract currently not available.

See more of this Session: CACS 30th Anniversary Symposium: Global Career Opportunities for Chemical Engineers
See more of this Group/Topical: Management Division

Development and Commercialization of Emulsion Aggregation Toner Technology by Xerox

2011 AIChE Annual Meeting
Minneapolis, MN
October 20th, 2011

By
Dr. Hadi Mahabadi
VP and Director, Xerox Research Centre of Canada

ABSTRACT

The Emulsion Aggreagation Toner Technology (EA) that was invented, developed and commercialized by Xerox Corporation represents a breakthrough in the chemistry and chemical engineering processes for prepararing toner materials using nanotechnology methodology. The innovation that uses a patented manufacturing scale semi-continuous emulsion polymerization and other proproietort processes to produce nano-scale toner ingridients with unique properties and the first-of-a-kind application of a nanotechnology methodology to generate micron-sized composite particles with tunable morphology, size and structure.

The technical challenges faced by chemical engineers and chemists during the early stages of product development will be presented.Since its commercialization, Xerox has launched three EA toner designs with the fourth generation on the horizon that will continue to deliver the wow factor to the customers and satisfy their unmet needs. The advantages of the technology that includes performance, economic, social and environmental benefits will be covered in this talk.

AUTHOR INDEX

AUTHOR INDEX

AUTHOR INDEX

AUTHOR INDEX

American Institute of Chemical Engineers
3 Park Avenue
New York, New York 10016-5991

CURRAN ASSOCIATES INC
proceedings.com